International Zinc Conference and International Zinc Oxide Industry Conference 2017

AA005452

Rancho Mirage, California, USA
26 February - 1 March 2017

ISBN: 978-1-5108-4062-1

Printed from e-media with permission by:

Curran Associates, Inc.
57 Morehouse Lane
Red Hook, NY 12571

Some format issues inherent in the e-media version may also appear in this print version.

Printed by Curran Associates, Inc. (2017)

For permission requests, please contact International Zinc Association
at the address below.

International Zinc Association
2530 Meridian Parkway
Suite 115
Durham, NC 27713
USA

Phone: +1 919 361 4647

contact@zinc.org

Additional copies of this publication are available from:

Curran Associates, Inc.
57 Morehouse Lane
Red Hook, NY 12571 USA
Phone: 845-758-0400
Fax: 845-758-2633
Email: curran@proceedings.com
Web: www.proceedings.com

TABLE OF CONTENTS

GENERAL SESSION

MORNING GENERAL SESSION

AFTERNOON SESSION JOINT WITH OXIDE INDUSTRY CONFERENCE

MORNING SESSION

AFTERNOON SESSION

ADDITIONAL PAPER

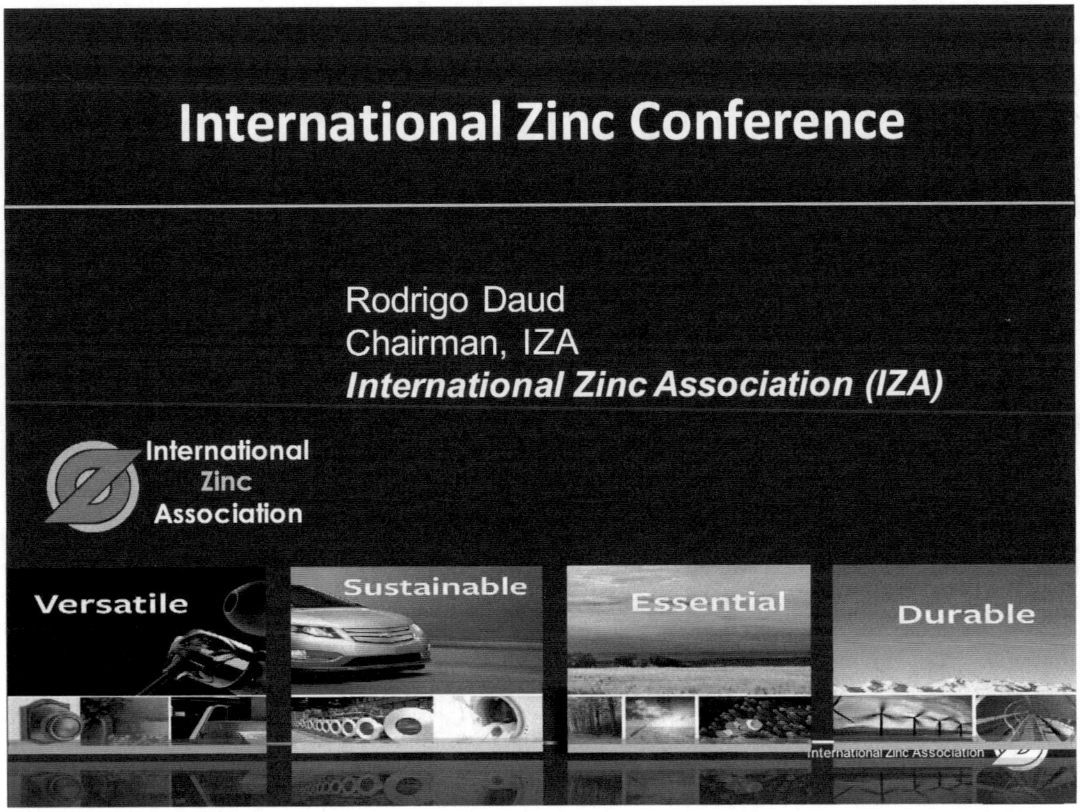
International Zinc Conference
Rodrigo Daud
Chairman, IZA
International Zinc Association (IZA)
International Zinc Association
Versatile
Sustainable
Essential
Durable
International Zinc Association

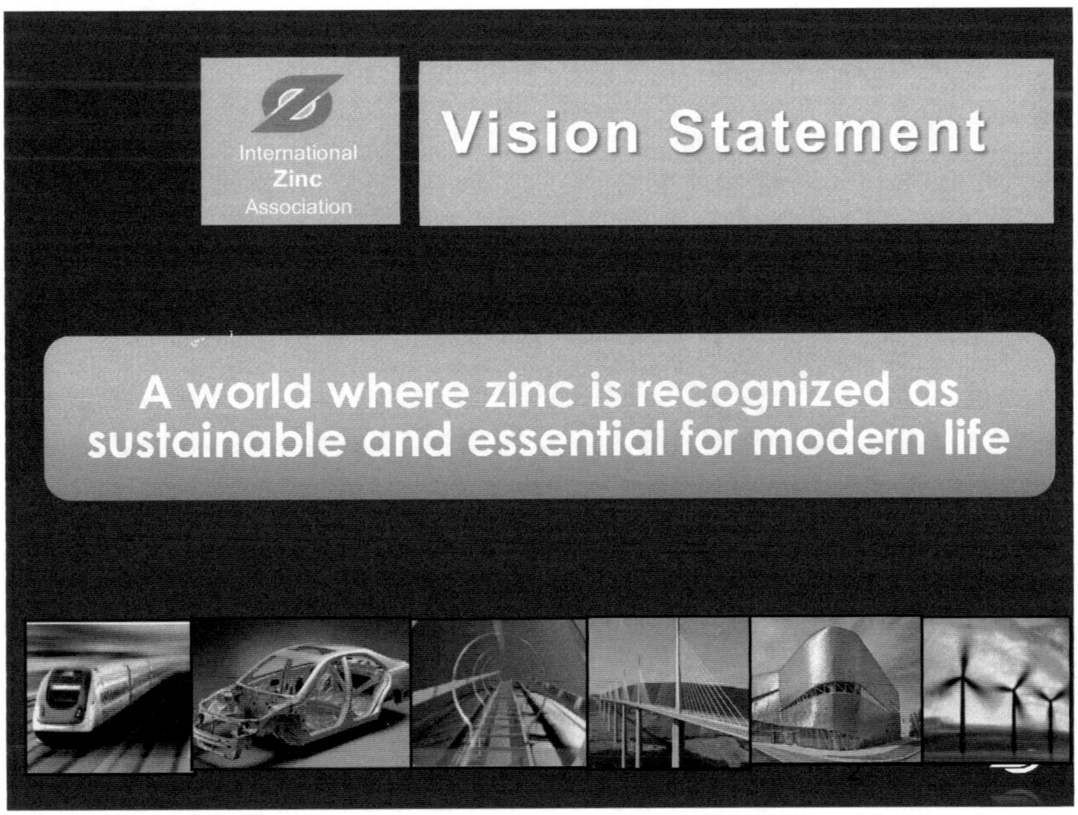
International Zinc Association
Vision Statement
A world where zinc is recognized as sustainable and essential for modern life

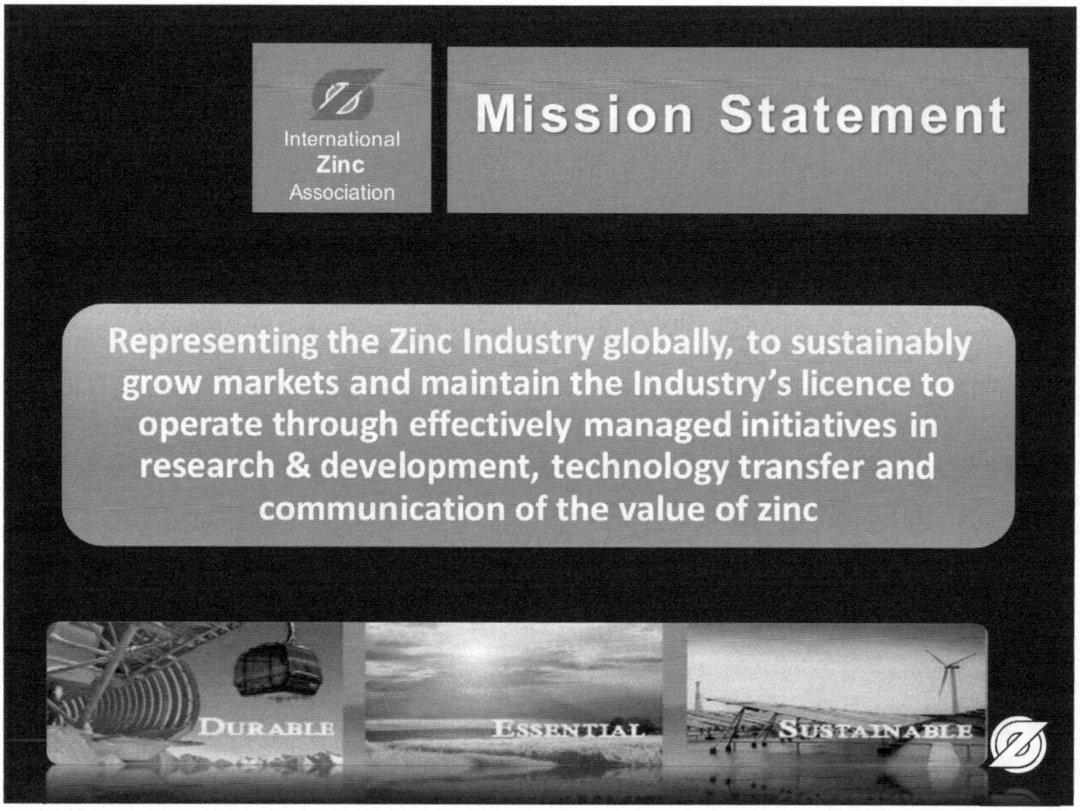
International
Zinc
Association
Mission Statement
Representing the Zinc Industry globally, to sustainably grow markets and maintain the Industry's licence to operate through effectively managed initiatives in research & development, technology transfer and communication of the value of zinc
DURABLE
ESSENTIAL
SUSTAINABLE

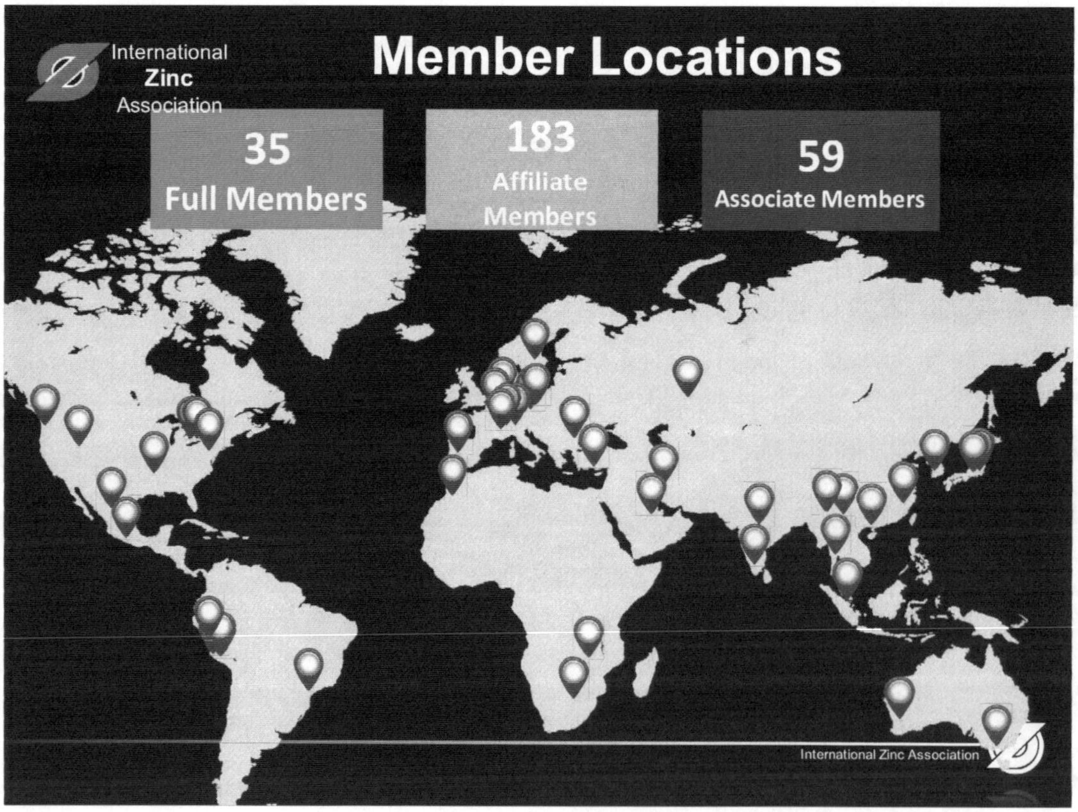
International
Zinc
Association
Member Locations
35
Full Members
183
Affiliate
Members
59
Associate Members
International Zinc Association

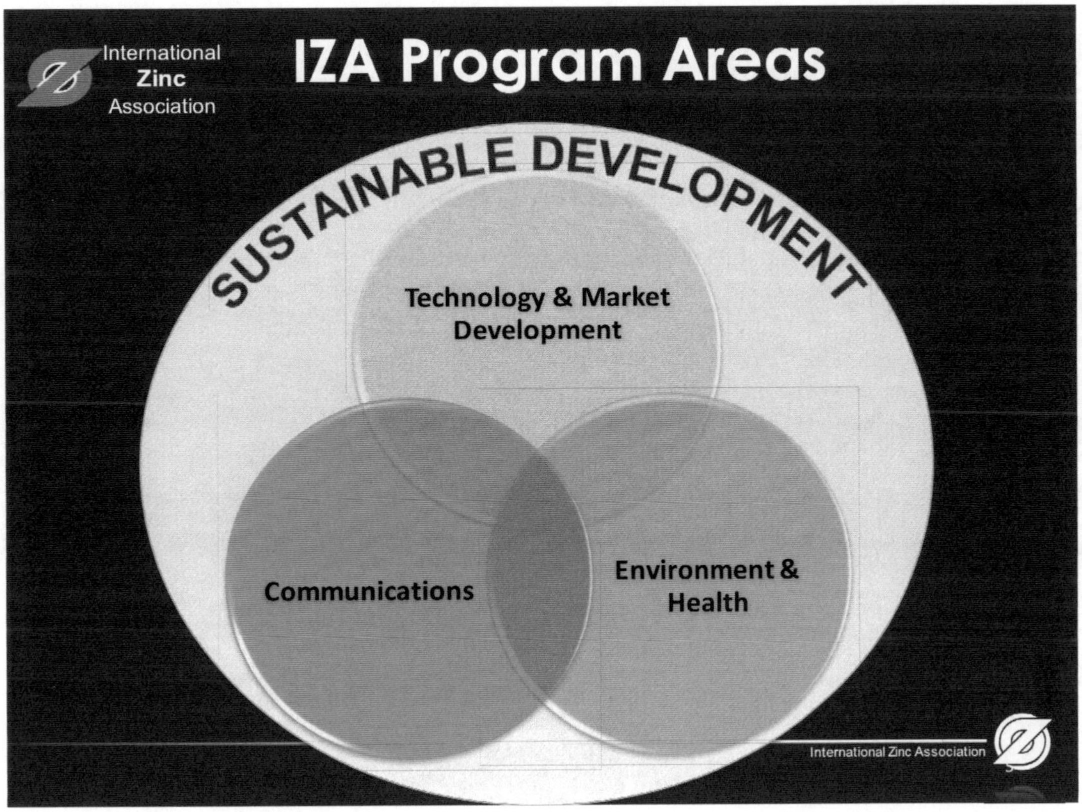
International Zinc Association
IZA Program Areas
SUSTAINABLE DEVELOPMENT
Technology & Market Development
Communications
Environment & Health
International Zinc Association

2017 International Zinc Conference - Europe
SAVE THE DATE!
May 8 -10, 2017
Where:
Steigenberger Wiltcher's
71 Avenue Lousie / Louizalaan 71
1050 Brussels, Belgium
International Zinc Association

Zinc College 2018
Location: Kokkola
BOLIDEN
Dates: June
Company Partner: Boliden
Field Tour: Boliden's zinc refinery,
Optional: mine tour
Kylylahti
BOLIDEN
Kokkola
SWEDEN

Zinc Video
ZINC
OWNABLE_
ORIGINAL_
CUSTOM_
BRANDED_
SOPHISTICATED_
PIONEERING_
INTRIGUING_
30
Zn
ZINC
ZINC
ESSENTIAL FOR MODERN LIFE

Zinc Video
ZINC
OWNABLE_
ORIGINAL_
CUSTOM_
BRANDED_
SOPHISTICATED_
PIONEERING_
INTRIGUING_
30
Zn
ZINC
ZINC
ESSENTIAL FOR MODERN LIFE

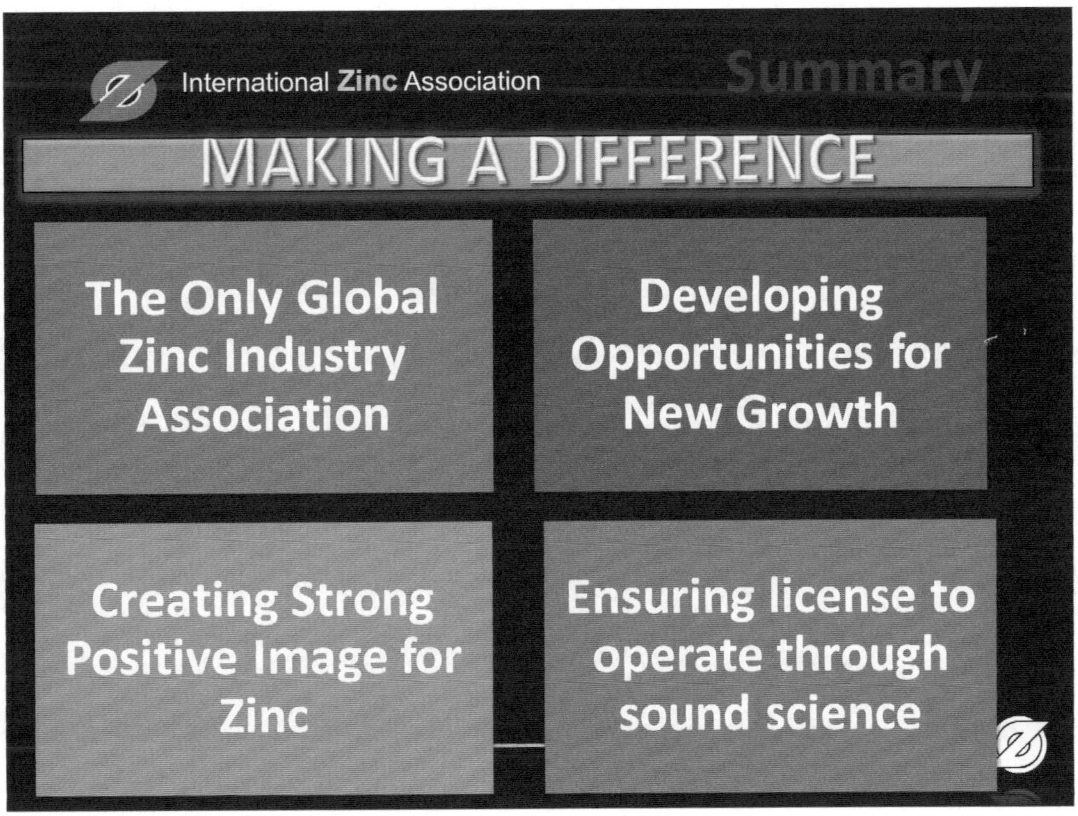
International Zinc Association
Summary
MAKING A DIFFERENCE
The Only Global Zinc Industry Association
Developing Opportunities for New Growth
Creating Strong Positive Image for Zinc
Ensuring license to operate through sound science

10 Global Flashpoints in 2017

In Libya with Militia during the Civil War -- 2012

With Moammar Qaddafi –Tripoli 1980

On the frontline with ISIS in Iraq

Perpetual disruption

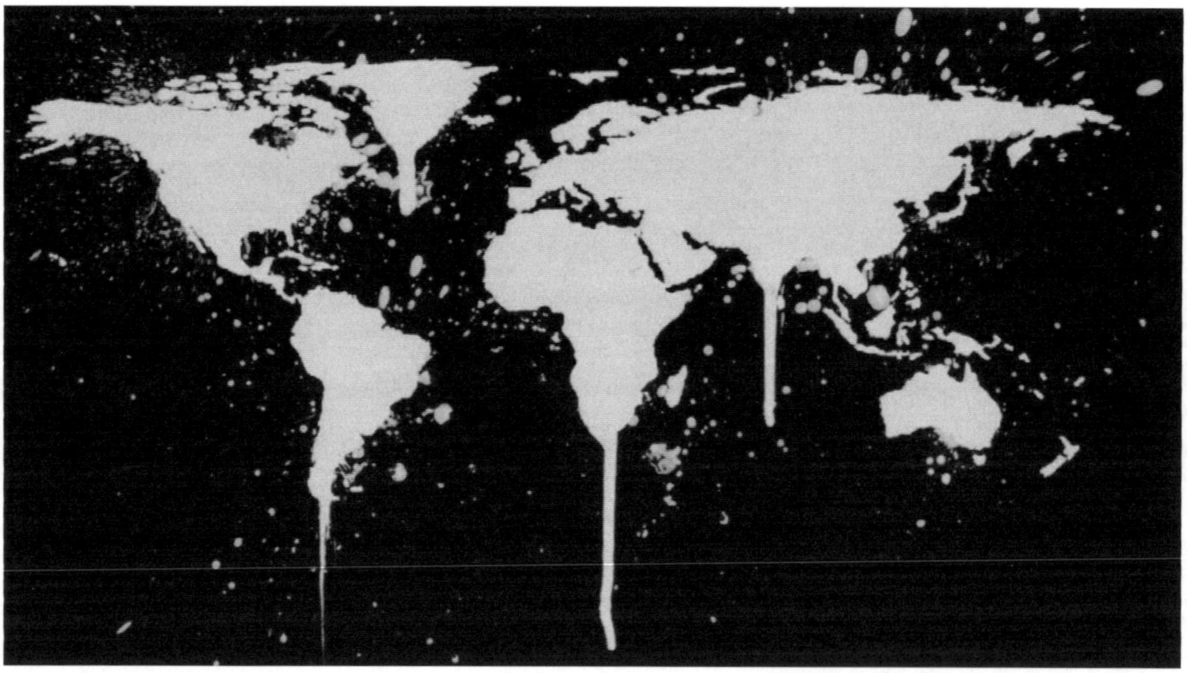

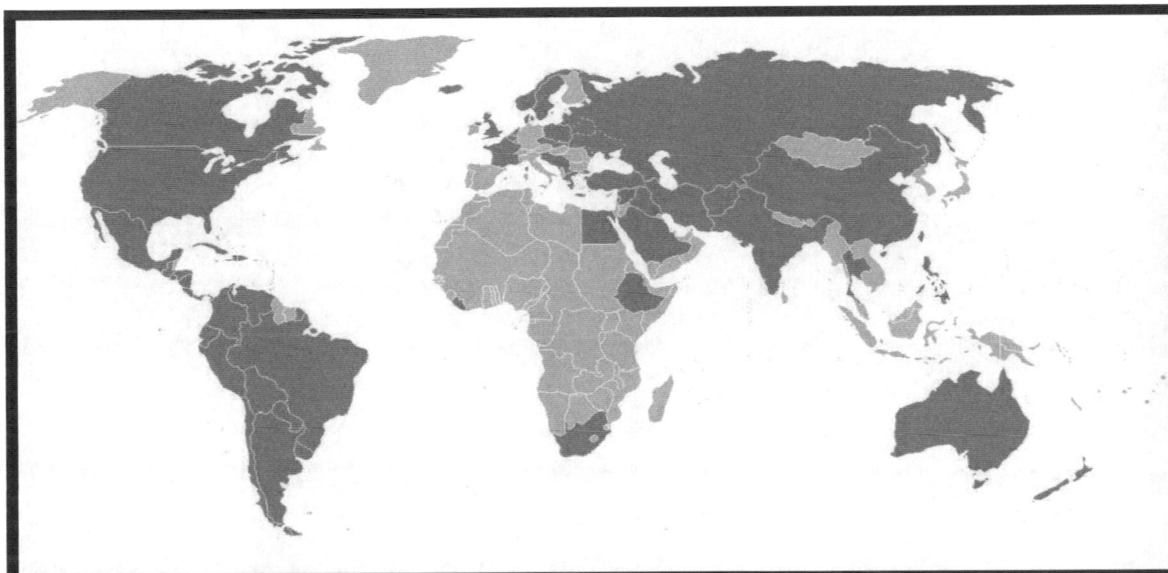

Original U.N. Membership - blue

U.N. Members today

Since 1990, 30 new countries in Europe, Asia, and Africa:

15 from the Soviet Union:

Armenia
Azerbaijan
Belarus
Estonia
Georgia
Kazakhstan
Kyrgyzstan
Latvia
Lithuania
Moldova
Russia
Tajikistan
Turkmenistan
Ukraine
Uzbekistan

7 from Yugoslavia in early 1990s

Bosnia and Herzegovina
Croatia
Macedonia
Montenegro
Slovenia
Serbia
Kosovo

13 Other changes:

- Namibia -- 1990
- North & South Yemen merged -- 1990
- E. Germany & W. Germany merged - 1990
- The Marshall Islands - 1991
- Micronesia – 1991
- Czech Republic and Slovakia -- 1993
- Eritrea seceded from Ethiopia - 1993
- Palau - 1994
- East Timor - 2002
- South Sudan - 2011

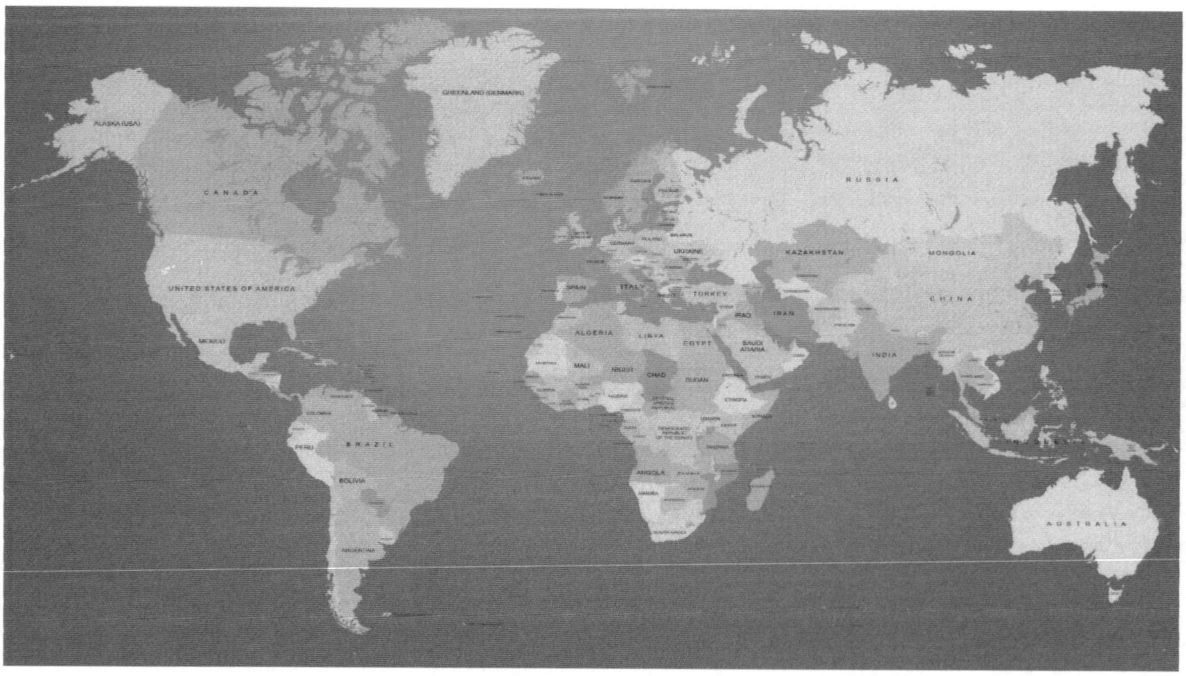

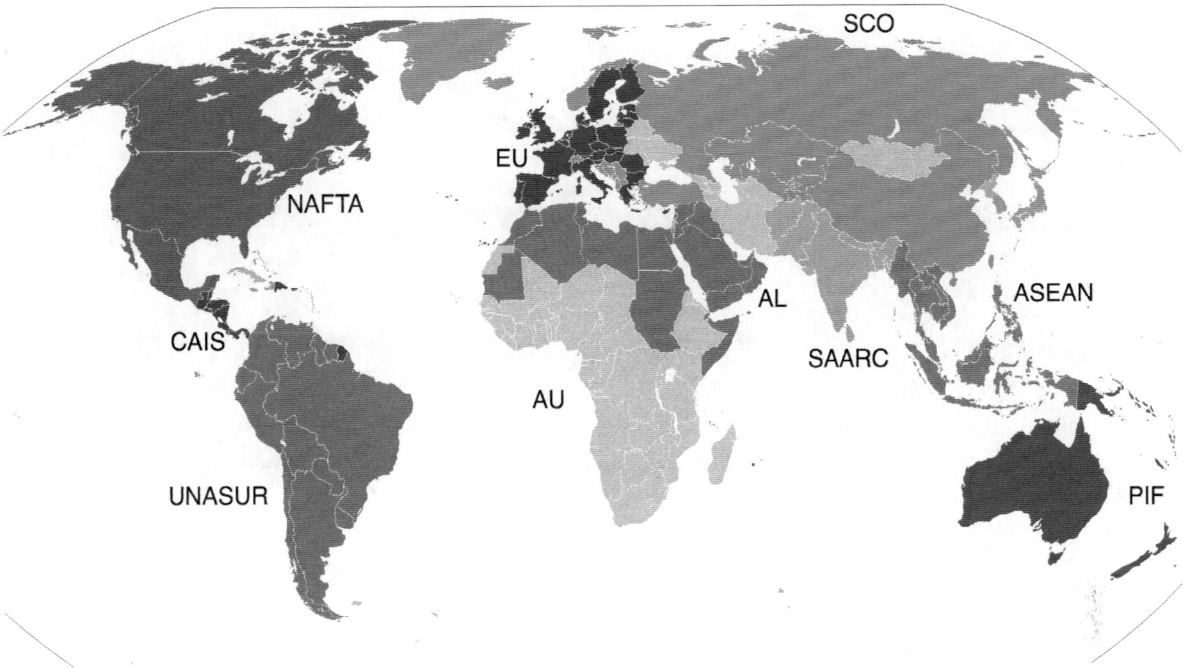
SCO
EU
NAFTA
AL
ASEAN
CAIS
SAARC
AU
UNASUR
PIF

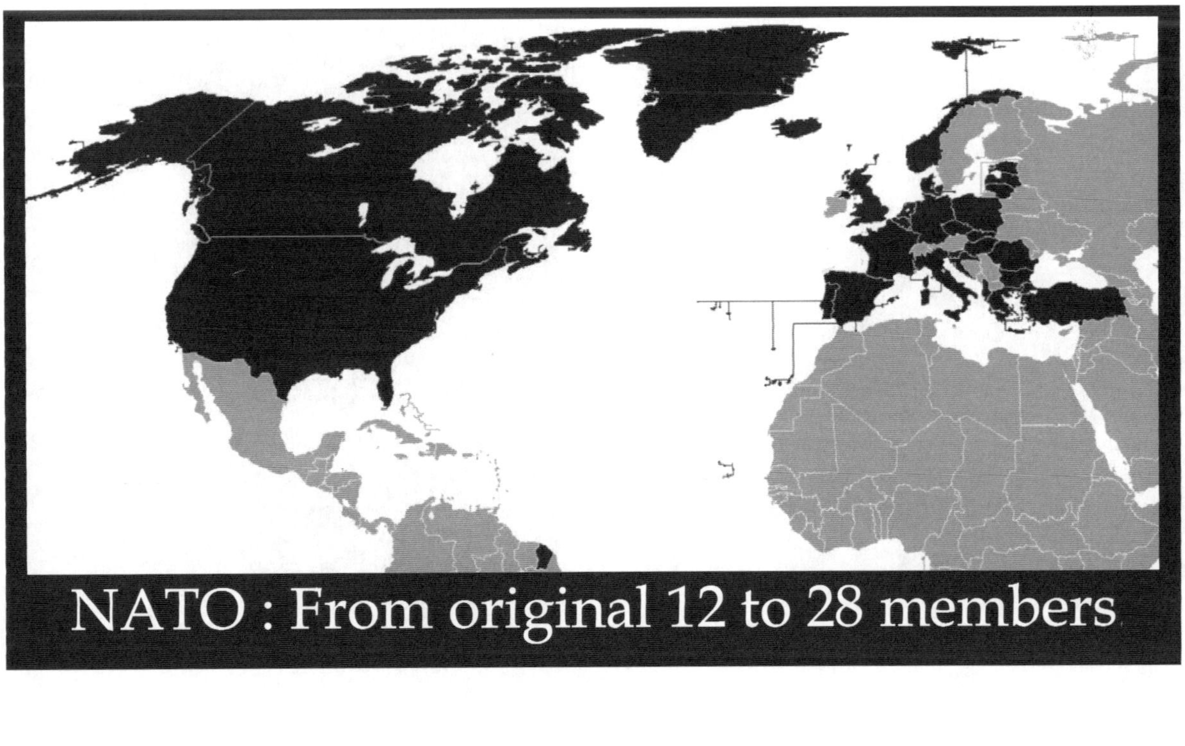
NATO : From original 12 to 28 members

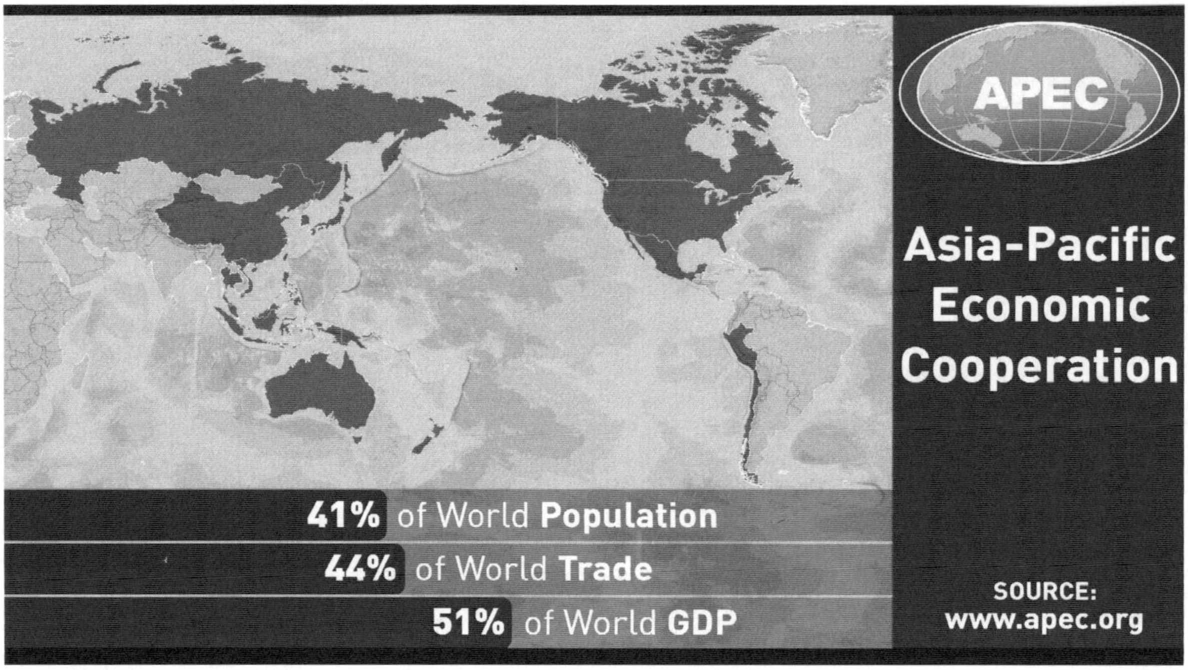
APEC
Asia-Pacific Economic Cooperation
41% of World Population
44% of World Trade
51% of World GDP
SOURCE: www.apec.org

MERCOSUR
14.869.755 km²
Español, Portugués y Guaraní
295.007.000
Argentina
Buenos Aires
3.761.274 km²
Español
41.775.000
IDH 0,81
Peso Argentino
Bolivia
Sucre (Capital Constitucional)
1.098.581 km²
Español y todos los idiomas de los naciones indígenas originarios campesinos
10.598.000
IDH 0,67
Boliviano
Referencias de Países
Estados Partes
Estados Asociados
Estado en proceso de adhesión
Brasil
Brasilia
8.514.880 km²
Portugués
201.497.000
IDH 0,73
Real
Paraguay
Asunción
406.750 km²
Español y Guaraní
6.888.000
IDH 0,66
Guaraní
Uruguay
Montevideo
176.200 km²
Español
3.418.000
IDH 0.79
Peso Uruguayo
Venezuela
Caracas
912.050 km²
Español
30.831.000
IDH 0,74
Bolívar
Andean Community & Mercusor

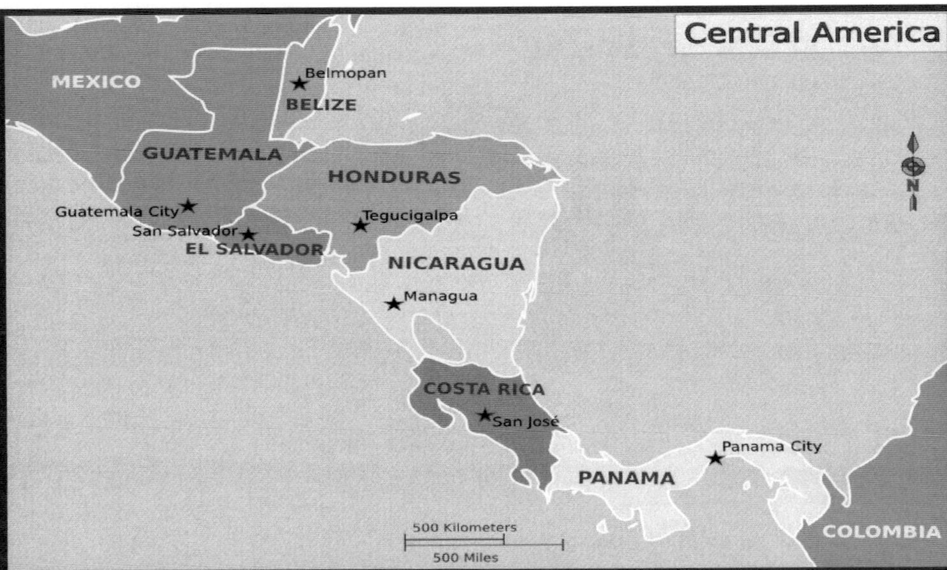

Central American Integration System

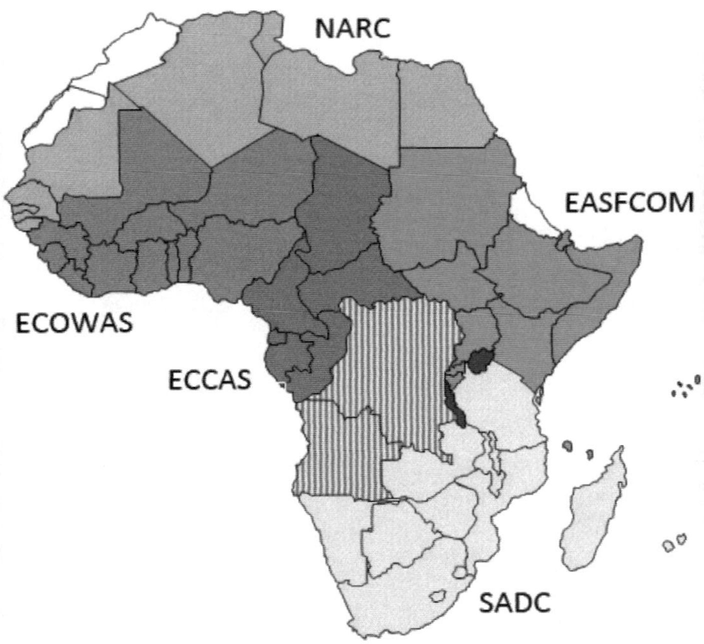

Economic & Military Alliances in Africa

"African Standby Forces"
Die in Afrika geplanten Eingreiftruppen
"North" NORTHBRIG (NASBRIG)
"West" ESF (ECOBRIG)
"Center" ECCAS (ECCASBRIG)
"East" EASBRIG
"South" SADCBRIG

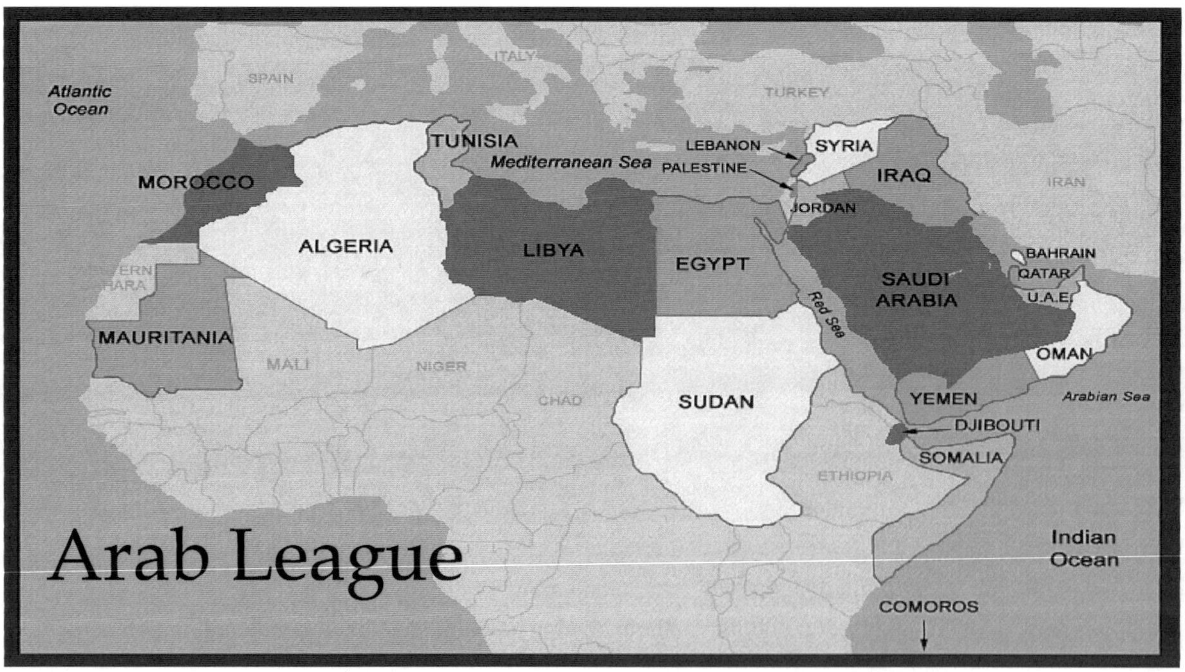
Atlantic Ocean
SPAIN
ITALY
TURKEY
TUNISIA
Mediterranean Sea
LEBANON
PALESTINE
SYRIA
IRAQ
IRAN
MOROCCO
JORDAN
ALGERIA
LIBYA
EGYPT
BAHRAIN
QATAR
U.A.E.
SAUDI ARABIA
Red Sea
MAURITANIA
MALI
NIGER
OMAN
CHAD
SUDAN
YEMEN
Arabian Sea
DJIBOUTI
SOMALIA
ETHIOPIA
Arab League
Indian Ocean
COMOROS

LEBANON
ISRAEL
IRAQ
IRAN
AFGHANISTAN
KUWAIT
PAKISTAN
GCC
BAHRAIN
QUATAR
SAUDI ARABIA
U.A.E
OMAN
INDIA
2015 © GreenOdin.com
Gulf Cooperation Council

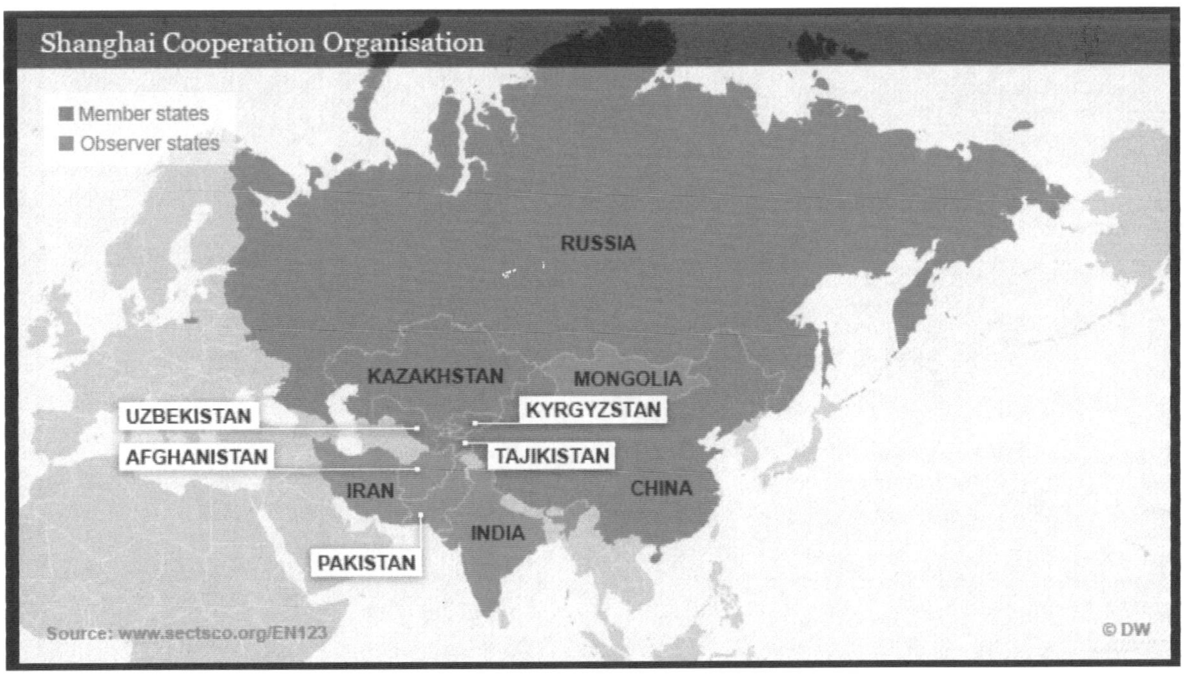
Shanghai Cooperation Organisation
Member states
Observer states
RUSSIA
KAZAKHSTAN
MONGOLIA
UZBEKISTAN
KYRGYZSTAN
AFGHANISTAN
TAJIKISTAN
IRAN
CHINA
INDIA
PAKISTAN
Source: www.sectsco.org/EN123
© DW

Commonwealth of Independent States

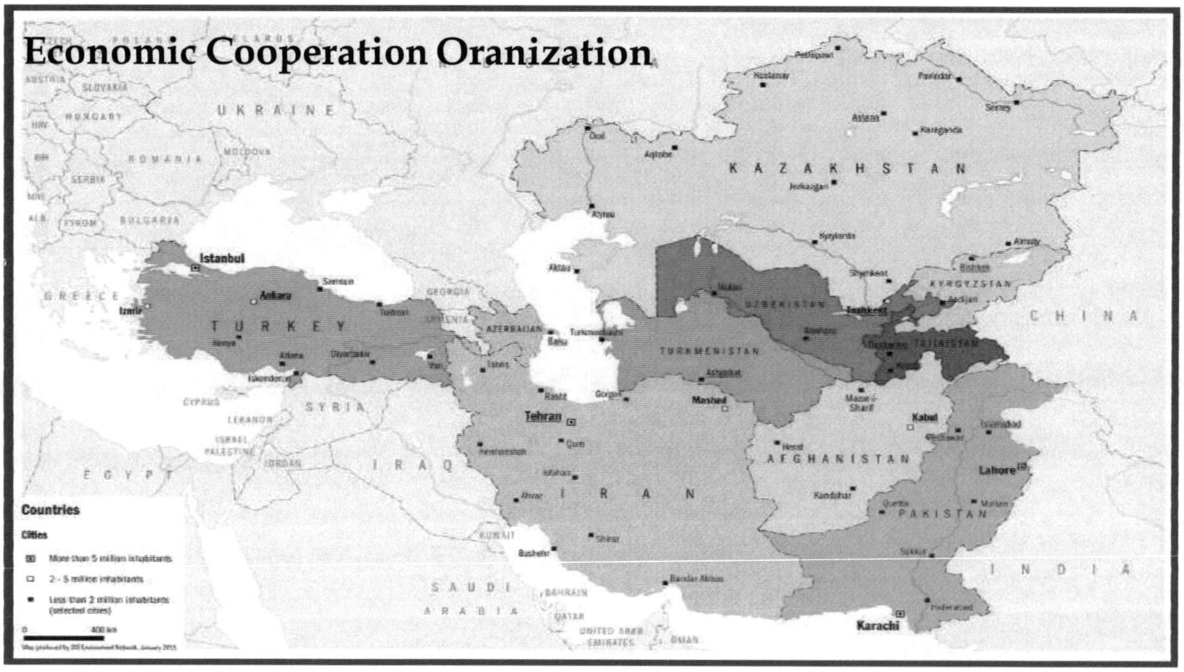

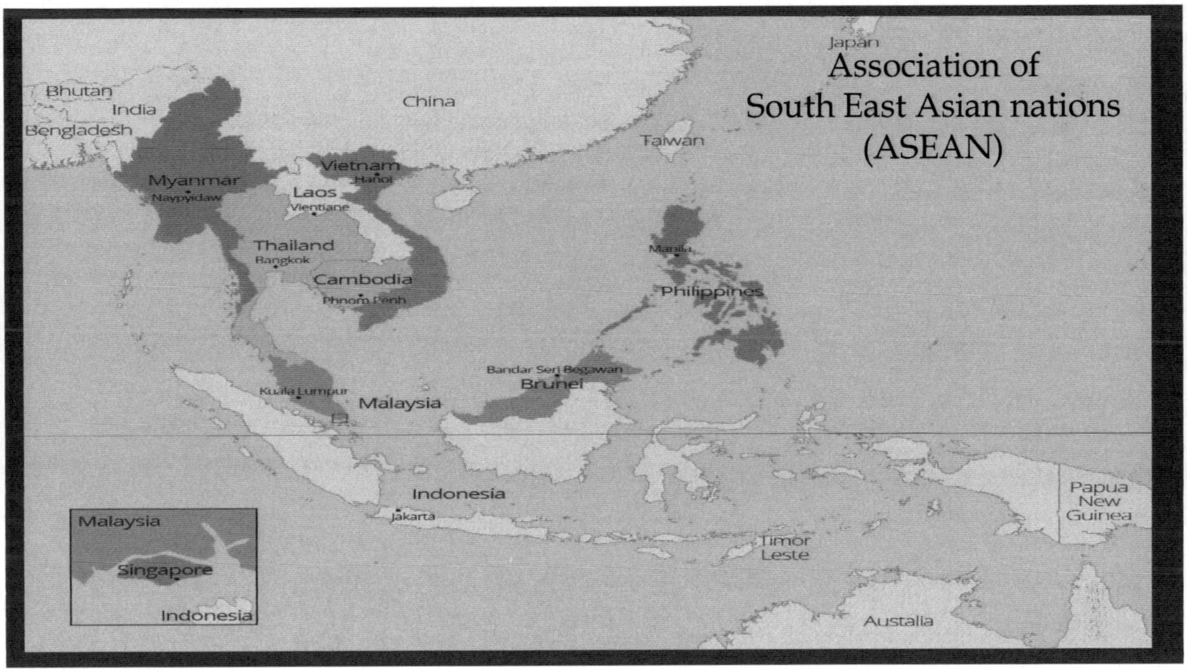
Association of
South East Asian nations
(ASEAN)
Japan
China
Bhutan
India
Bengladesh
Myanmar
Naypyidaw
Vietnam
Hanoi
Laos
Vientiane
Thailand
Bangkok
Cambodia
Phnom Penh
Taiwan
Manila
Philippines
Bandar Seri Begawan
Brunei
Kuala Lumpur
Malaysia
Indonesia
Jakarta
Papua
New
Guinea
Timor
Leste
Austalia
Malaysia
Singapore
Indonesia

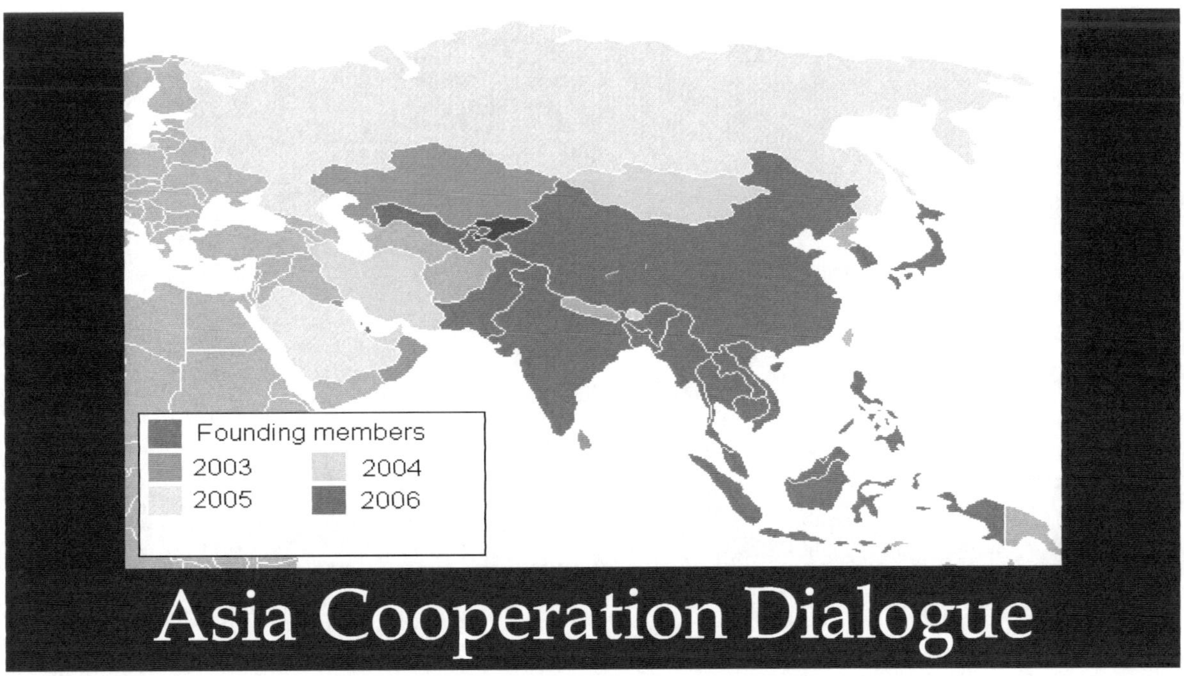
Founding members
2003
2004
2005
2006
Asia Cooperation Dialogue

Asian Cooperation Dialogue

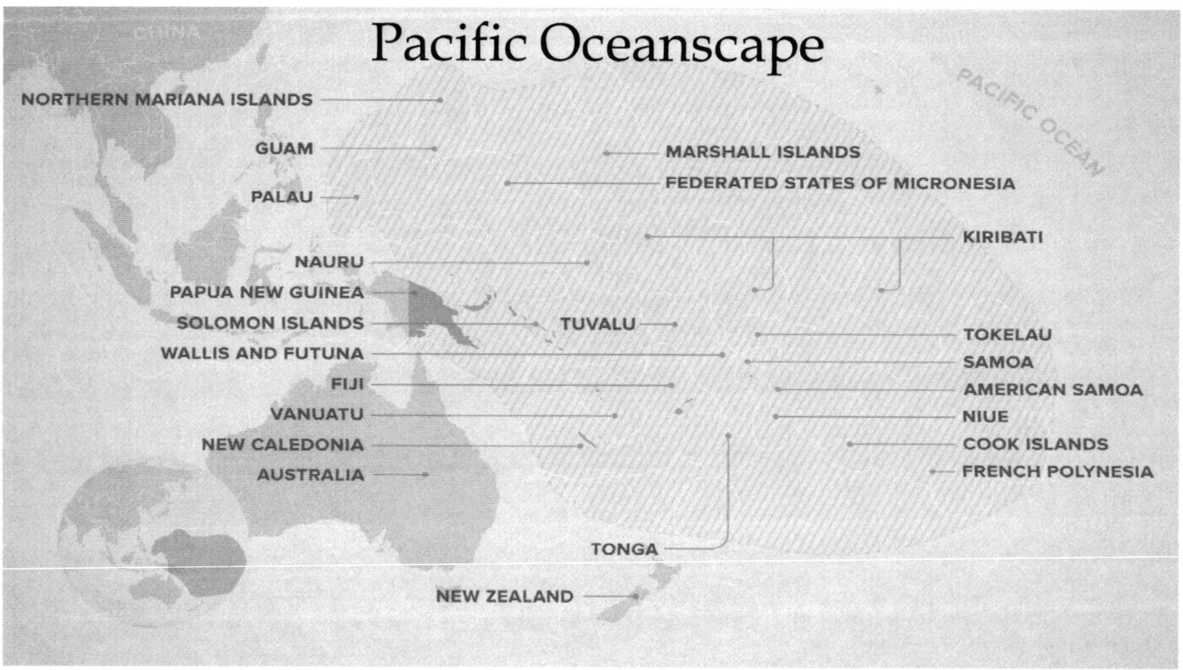
Pacific Oceanscape
PACIFIC OCEAN
NORTHERN MARIANA ISLANDS
GUAM
PALAU
MARSHALL ISLANDS
FEDERATED STATES OF MICRONESIA
KIRIBATI
NAURU
PAPUA NEW GUINEA
SOLOMON ISLANDS
TUVALU
WALLIS AND FUTUNA
FIJI
VANUATU
NEW CALEDONIA
AUSTRALIA
TOKELAU
SAMOA
AMERICAN SAMOA
NIUE
COOK ISLANDS
FRENCH POLYNESIA
TONGA
NEW ZEALAND

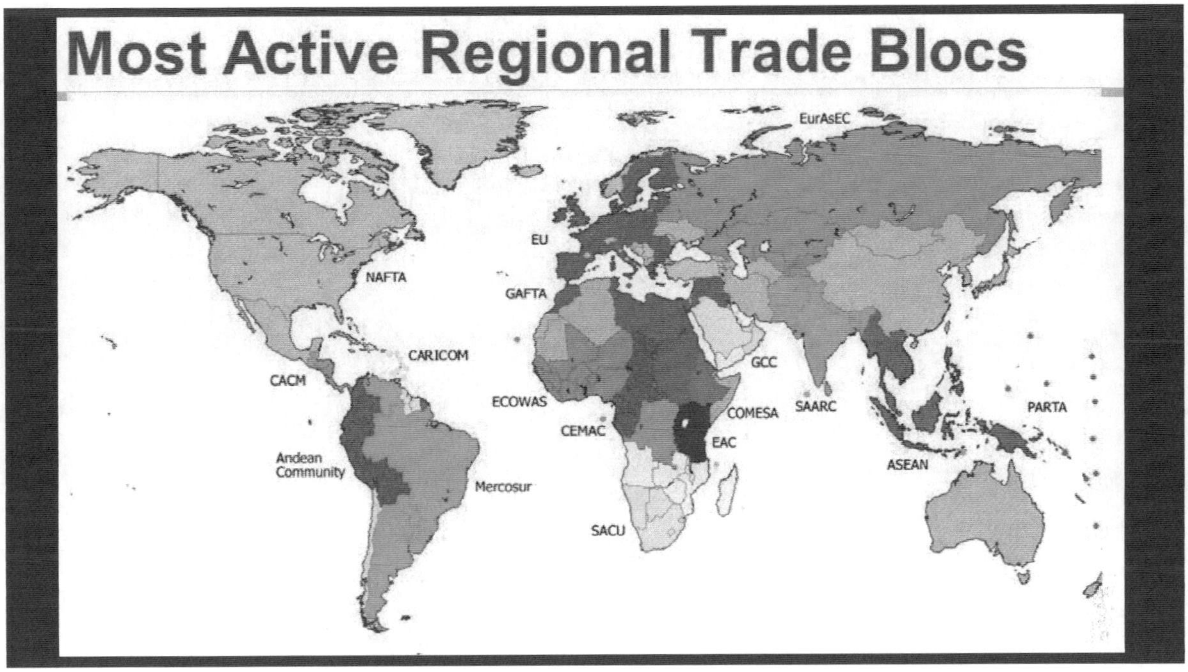
Most Active Regional Trade Blocs
EurAsEC
EU
NAFTA
GAFTA
CARICOM
GCC
CACM
ECOWAS
COMESA
SAARC
PARTA
CEMAC
EAC
Andean Community
ASEAN
Mercosur
SACU

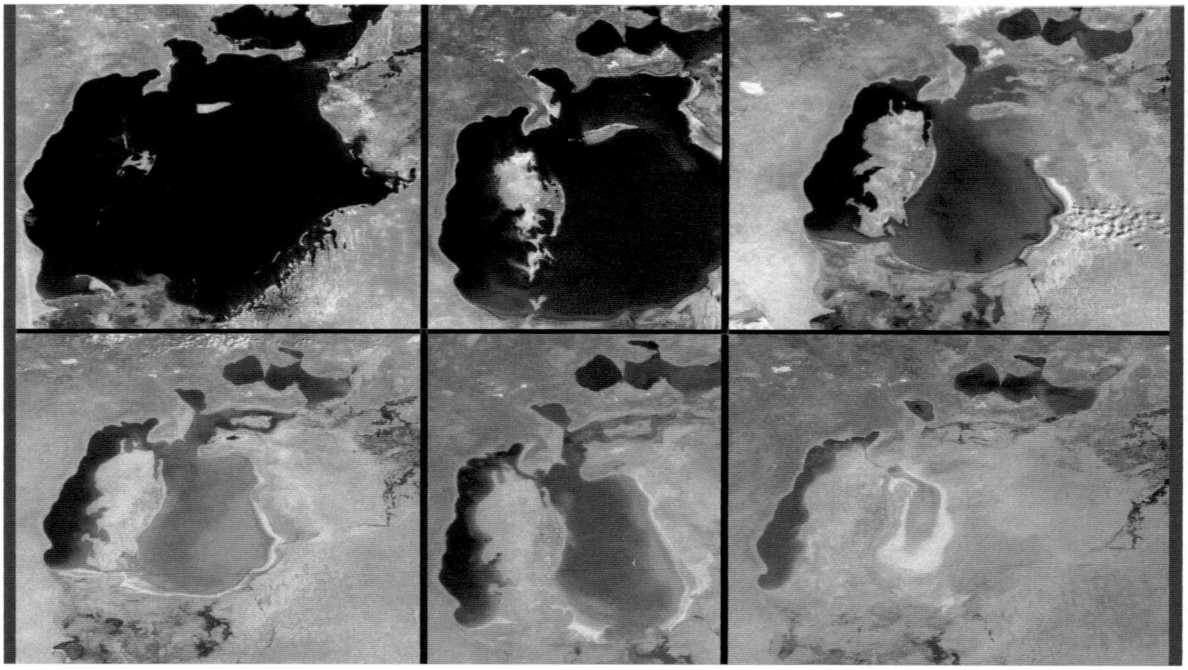

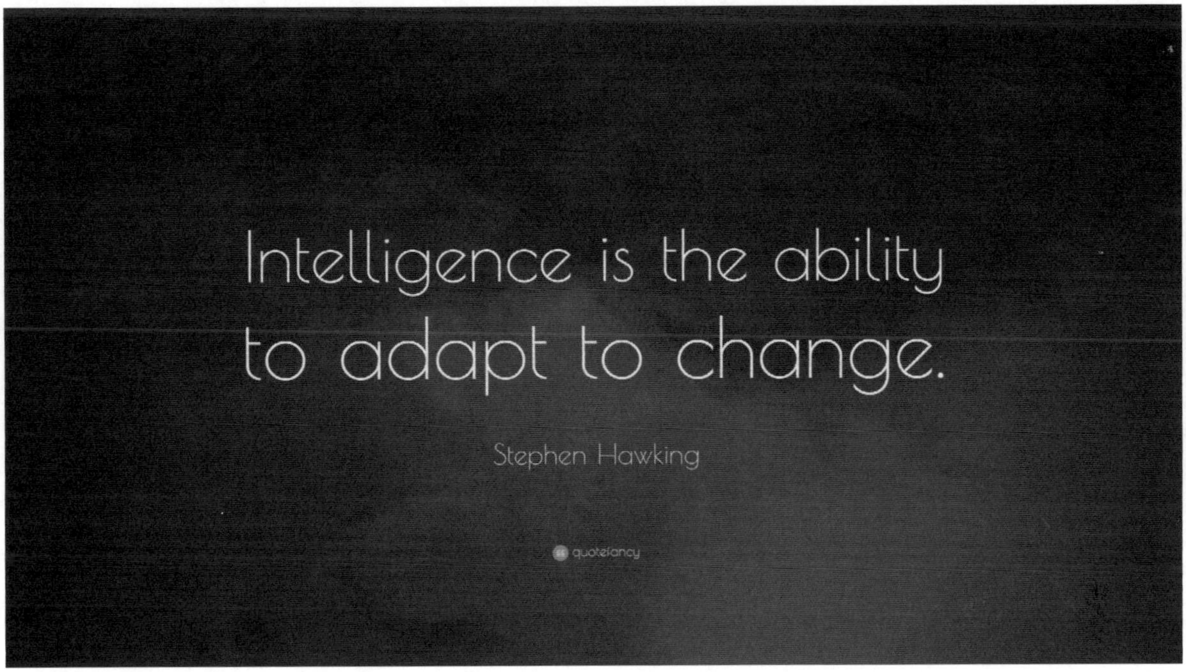
Intelligence is the ability
to adapt to change.
Stephen Hawking
quotefancy

© AFP/Getty Images

MEAN WORLD SYNDROME

NUMBER OF ARMED CONFLICTS AND WARS
1946-2014

Source: Uppsala Data Conflict Program/Peace Research Institute Oslo Armed Conflict Dataset.

RATE OF BATTLE DEATHS IN ARMED CONFLICTS
1946-2013

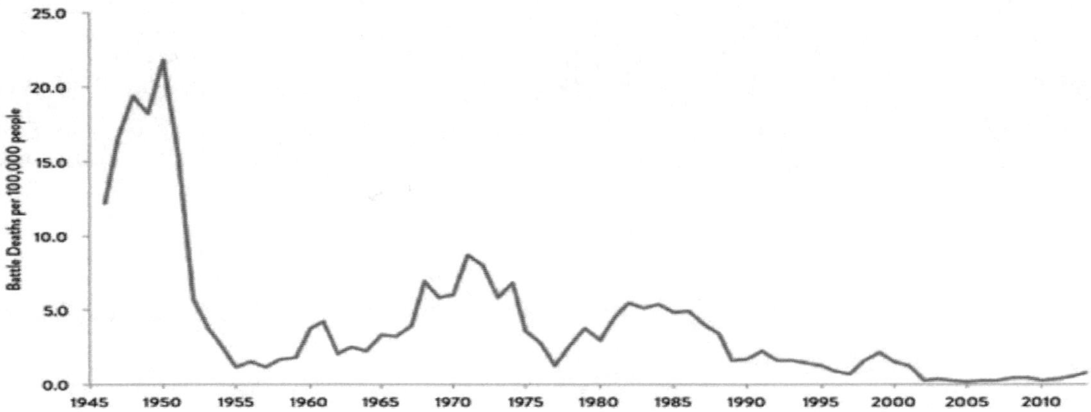

Source: Uppsala Conflict Data Program Battle-Related Deaths Dataset v.5-2014, 1989-2013, "Best Estimates." Missing estimate for 2013 for the Syrian civil war calculated from a revised UCDP "Low Estimate" and a "High Estimate" from the Peace Research Institute Oslo (both obtained in consultation with Erik Melander of UCDP). The "High Estimate" is for May 2013–April 2014, rather than calendar year 2013, and comes from Page 9 of M. Price, A. Gohdes, & P. Ball, "Updated Statistical Analysis of Documentation of Killings in the Syrian Arab Republic," Human Rights Data Analysis Group, http://www.ohchr.org/Documents/Countries/SY/HRDAGUpdatedReportAug2014.pdf. World population figures from U.S. Census Bureau, http://www.census.gov/population/international/data/worldpop/-table_population.php

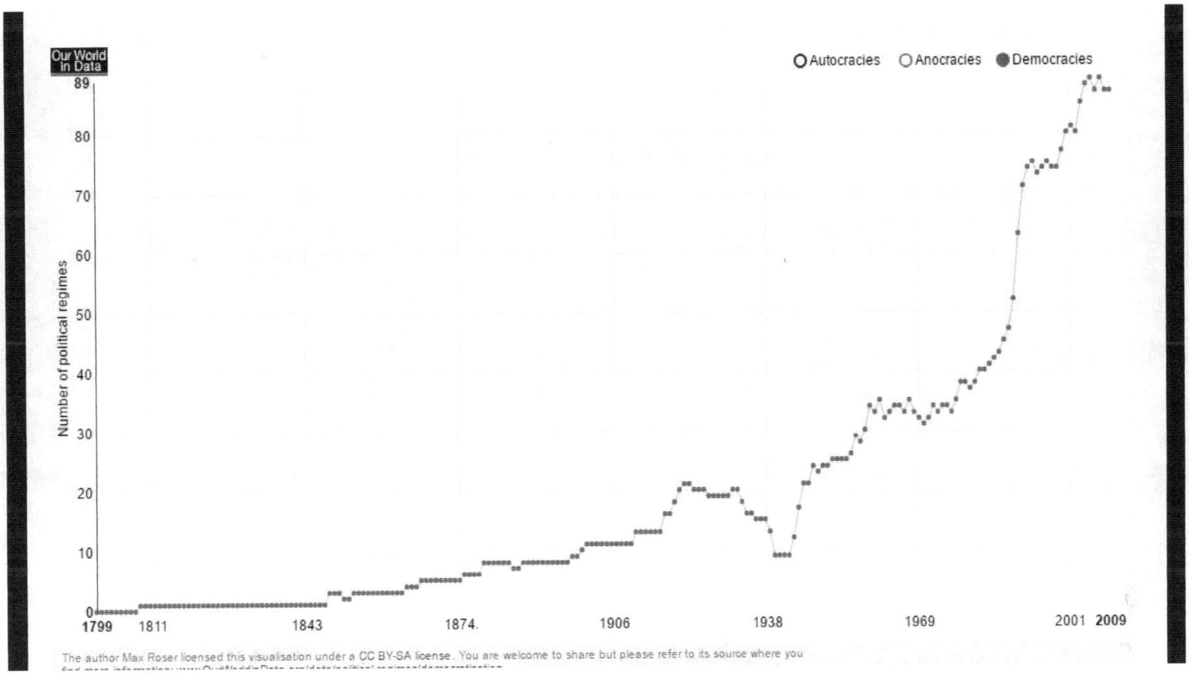
Our World in Data
Autocracies
Anocracies
Democracies
Number of political regimes
89
80
70
60
50
40
30
20
10
0
1799
1811
1843
1874
1906
1938
1969
2001
2009
The author Max Roser licensed this visualisation under a CC BY-SA license. You are welcome to share but please refer to its source where you

DEMOCRACY AND AUTOCRACY
1946-2013
Sum of Democracy or Autocracy Scores of Independent States with Population > 500,000
1000
900
800
700
600
500
400
300
200
100
0
1945
1950
1955
1960
1965
1970
1975
1980
1985
1990
1995
2000
2005
2010
DEMOCRACY SCORE
AUTOCRACY SCORE
Source: Polity IV Project, Center for Systemic Peace
(http://www.systemicpeace.org/polityproject.html),
"Global Democracy and Autocracy, 1946-2013."

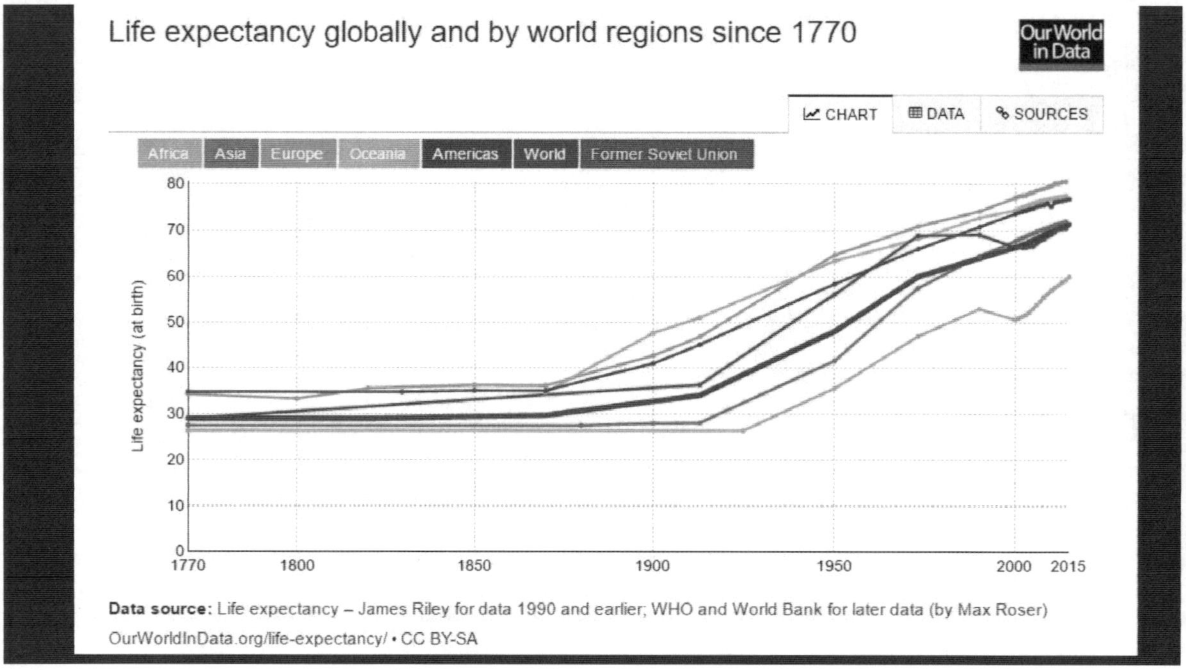
Life expectancy globally and by world regions since 1770
Our World in Data
CHART
DATA
SOURCES
Africa
Asia
Europe
Oceania
Americas
World
Former Soviet Union
Life expectancy (at birth)
80
70
60
50
40
30
20
10
0
1770
1800
1850
1900
1950
2000
2015
Data source: Life expectancy – James Riley for data 1990 and earlier; WHO and World Bank for later data (by Max Roser)
OurWorldInData.org/life-expectancy/ • CC BY-SA

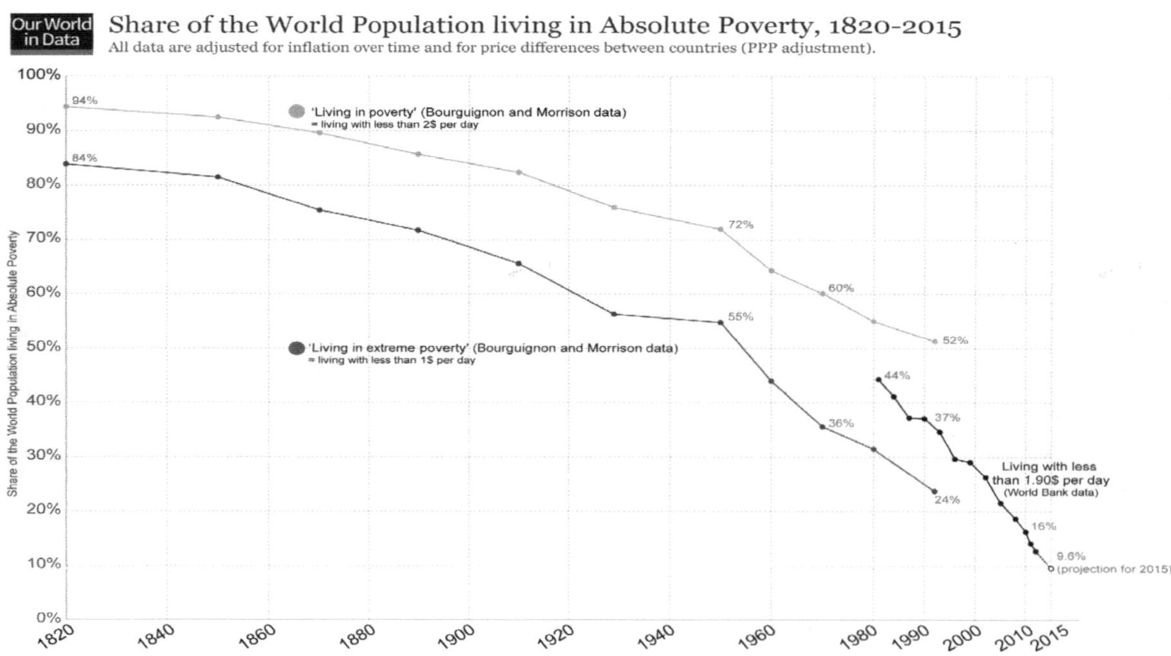
Our World in Data
Share of the World Population living in Absolute Poverty, 1820-2015
All data are adjusted for inflation over time and for price differences between countries (PPP adjustment).
Share of the World Population living in Absolute Poverty
100%
90%
80%
70%
60%
50%
40%
30%
20%
10%
0%
'Living in poverty' (Bourguignon and Morrison data)
= living with less than 2$ per day
'Living in extreme poverty' (Bourguignon and Morrison data)
= living with less than 1$ per day
Living with less than 1.90$ per day
(World Bank data)
94%
84%
72%
60%
55%
52%
44%
37%
36%
24%
16%
9.6%
(projection for 2015)
1820
1840
1860
1880
1900
1920
1940
1960
1980
1990
2000
2010
2015

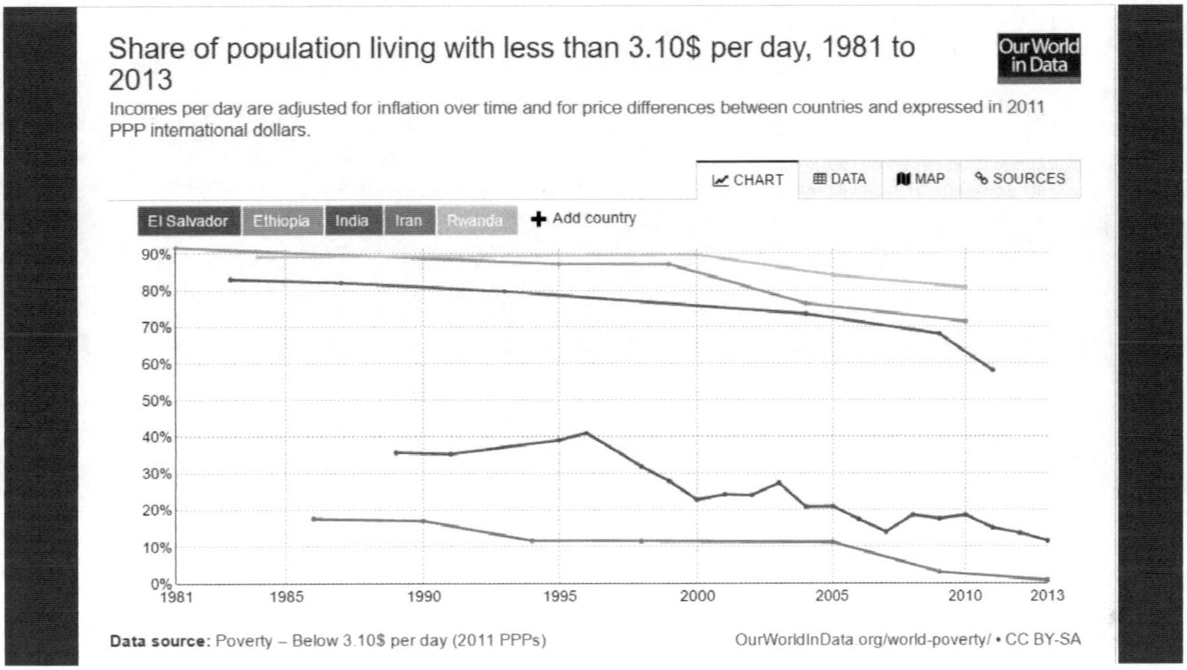
Share of population living with less than 3.10$ per day, 1981 to 2013
Our World in Data
Incomes per day are adjusted for inflation over time and for price differences between countries and expressed in 2011 PPP international dollars.
CHART
DATA
MAP
SOURCES
El Salvador
Ethiopia
India
Iran
Rwanda
Add country
90%
80%
70%
60%
50%
40%
30%
20%
10%
0%
1981
1985
1990
1995
2000
2005
2010
2013
Data source: Poverty – Below 3.10$ per day (2011 PPPs)
OurWorldInData.org/world-poverty/ • CC BY-SA

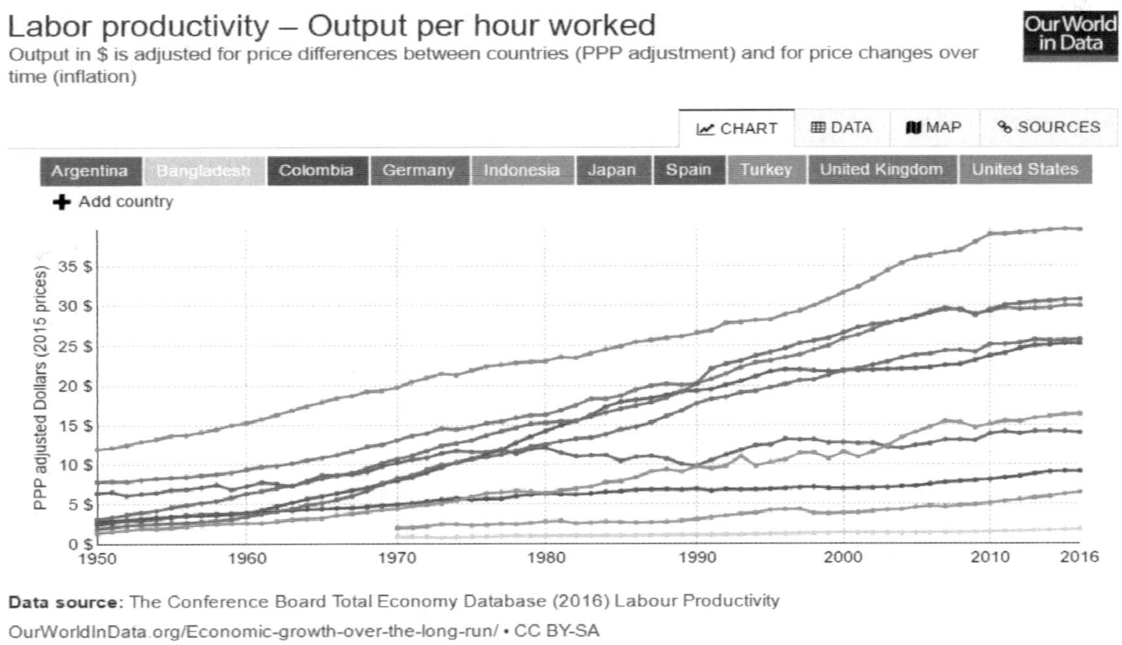
Labor productivity – Output per hour worked
Our World in Data
Output in $ is adjusted for price differences between countries (PPP adjustment) and for price changes over time (inflation)
CHART
DATA
MAP
SOURCES
Argentina
Bangladesh
Colombia
Germany
Indonesia
Japan
Spain
Turkey
United Kingdom
United States
Add country
PPP adjusted Dollars (2015 prices)
35 $
30 $
25 $
20 $
15 $
10 $
5 $
0 $
1950
1960
1970
1980
1990
2000
2010
2016
Data source: The Conference Board Total Economy Database (2016) Labour Productivity
OurWorldInData.org/Economic-growth-over-the-long-run/ • CC BY-SA

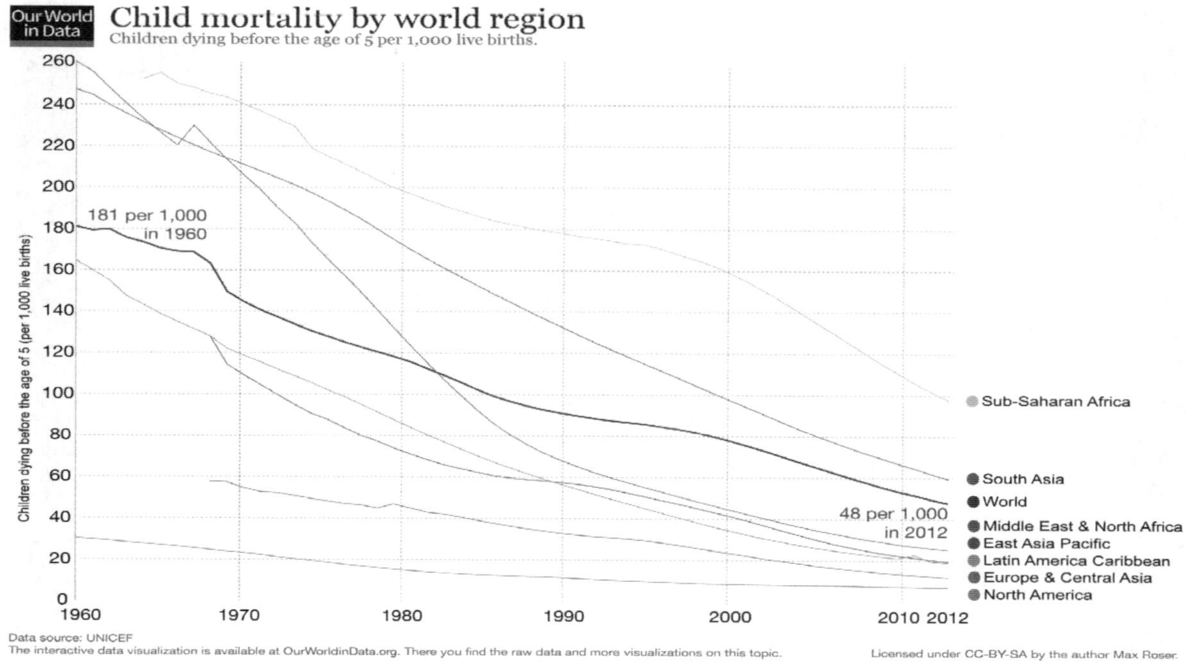
Our World in Data
Child mortality by world region
Children dying before the age of 5 per 1,000 live births.
Children dying before the age of 5 (per 1,000 live births)
181 per 1,000 in 1960
48 per 1,000 in 2012
Sub-Saharan Africa
South Asia
World
Middle East & North Africa
East Asia Pacific
Latin America Caribbean
Europe & Central Asia
North America
Data source: UNICEF
The interactive data visualization is available at OurWorldinData.org. There you find the raw data and more visualizations on this topic.
Licensed under CC-BY-SA by the author Max Roser.

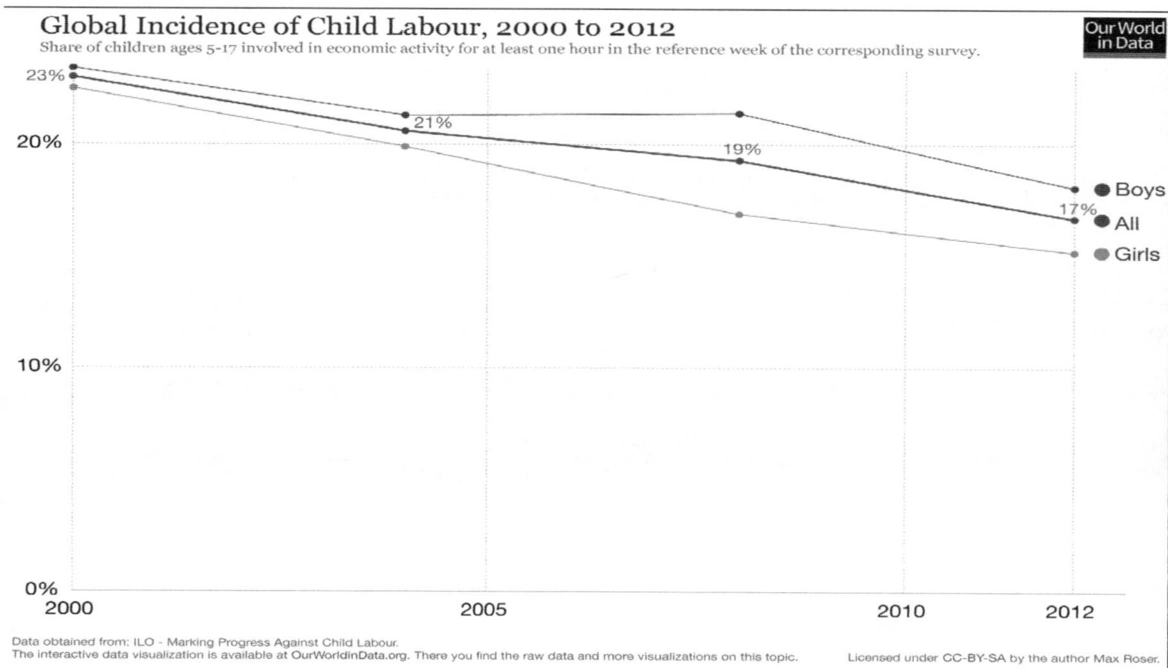
Global Incidence of Child Labour, 2000 to 2012
Share of children ages 5-17 involved in economic activity for at least one hour in the reference week of the corresponding survey.
Our World in Data
23%
21%
19%
17%
Boys
All
Girls
Data obtained from: ILO - Marking Progress Against Child Labour.
The interactive data visualization is available at OurWorldinData.org. There you find the raw data and more visualizations on this topic.
Licensed under CC-BY-SA by the author Max Roser.

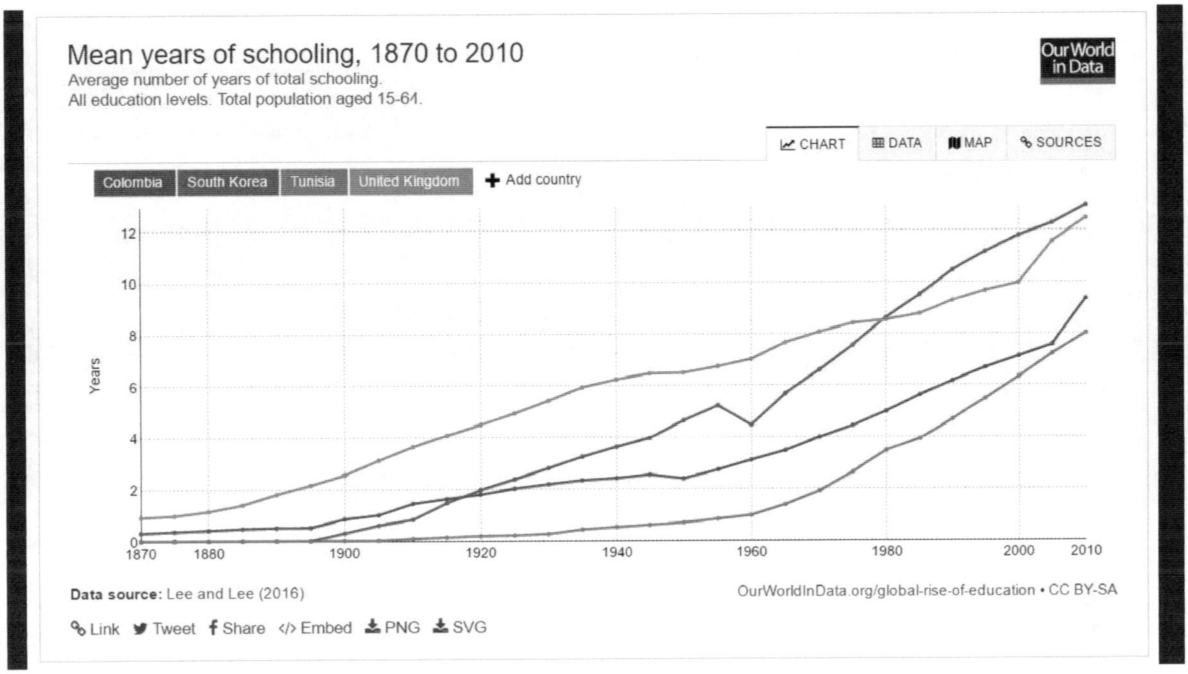
Mean years of schooling, 1870 to 2010
Average number of years of total schooling.
All education levels. Total population aged 15-64.
Our World in Data
CHART
DATA
MAP
SOURCES
Colombia
South Korea
Tunisia
United Kingdom
Add country
Years
0
2
4
6
8
10
12
1870
1880
1900
1920
1940
1960
1980
2000
2010
Data source: Lee and Lee (2016)
OurWorldInData.org/global-rise-of-education • CC BY-SA
Link
Tweet
Share
Embed
PNG
SVG

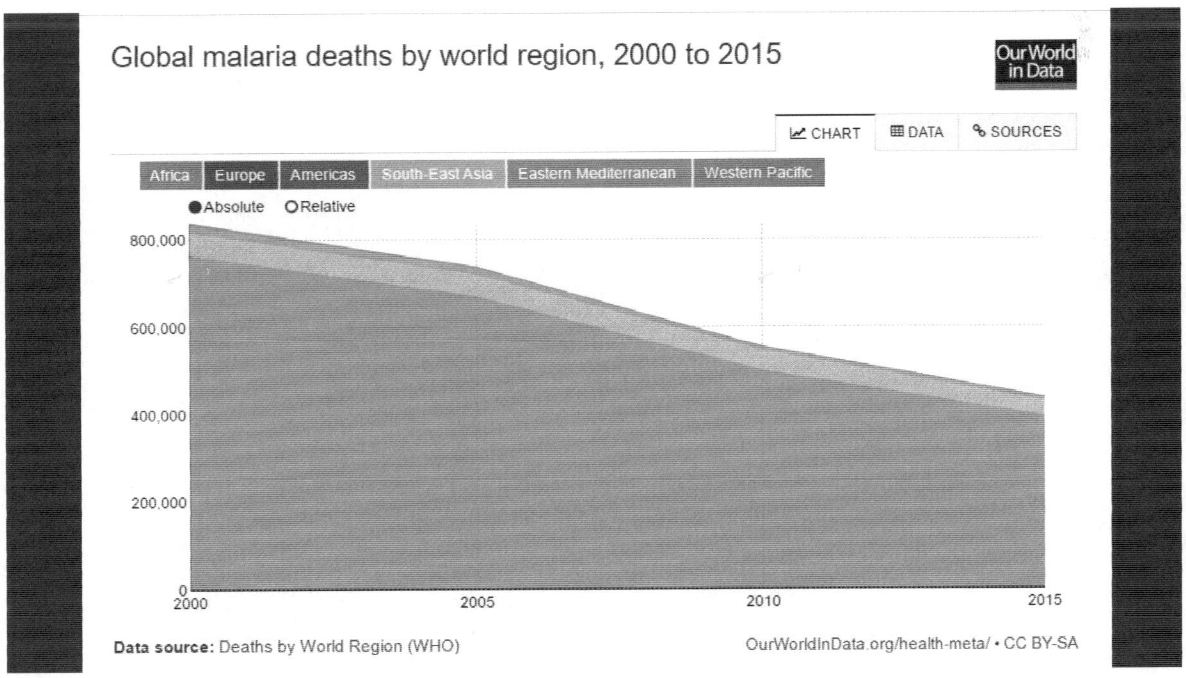
Global malaria deaths by world region, 2000 to 2015
Our World in Data
CHART
DATA
SOURCES
Africa
Europe
Americas
South-East Asia
Eastern Mediterranean
Western Pacific
Absolute
Relative
0
200,000
400,000
600,000
800,000
2000
2005
2010
2015
Data source: Deaths by World Region (WHO)
OurWorldInData.org/health-meta/ • CC BY-SA

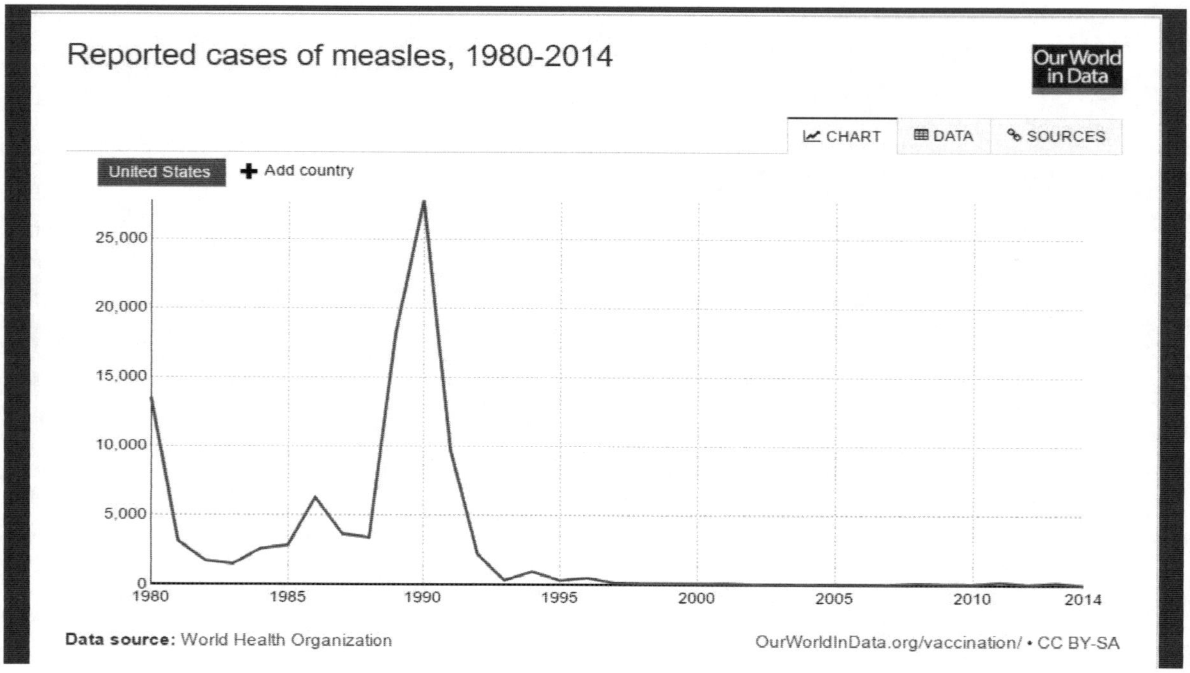
Reported cases of measles, 1980-2014
Our World in Data
CHART
DATA
SOURCES
United States
Add country
25,000
20,000
15,000
10,000
5,000
0
1980
1985
1990
1995
2000
2005
2010
2014
Data source: World Health Organization
OurWorldInData.org/vaccination/ • CC BY-SA

HOMICIDE RATES IN THE US AND ENGLAND
1967-2013, AND THE WORLD, 2003-2012
UNITED STATES
WORLD
ENGLAND
Homicides per 100,000 people
10
9
8
7
6
5
4
3
2
1
0
1965
1970
1975
1980
1985
1990
1995
2000
2005
2010
Sources: U.S., FBI Uniform Crime Reports. England (including Wales): U.K. Office for National Statistics. World: U.N. Office on Drugs and Crime, reported in U.N. Economic and Social Council's "World crime trends and emerging issues and responses in the field of crime prevention and social justice," Feb. 12, 2014, Figure 1. The percentages were converted to homicide rates by setting the 2012 rate at 6.2, the figure reported in the UNODC Global Study on Homicide 2013, Page 12.

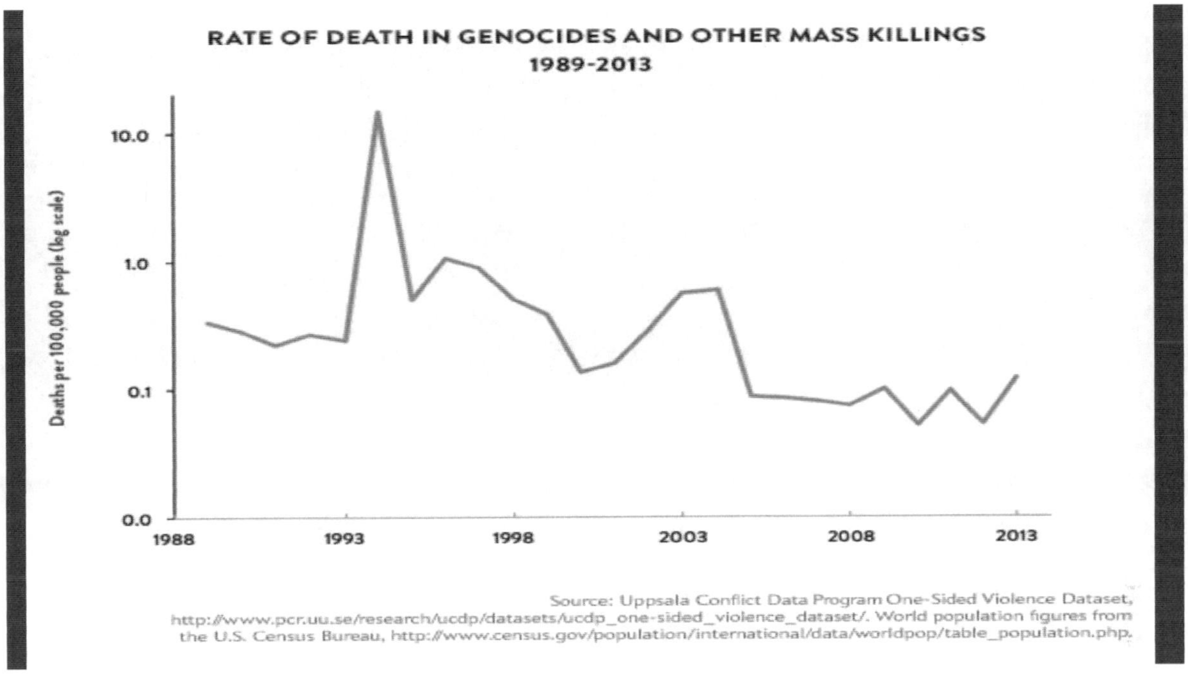
RATE OF DEATH IN GENOCIDES AND OTHER MASS KILLINGS
1989-2013
Deaths per 100,000 people (log scale)
10.0
1.0
0.1
0.0
1988
1993
1998
2003
2008
2013
Source: Uppsala Conflict Data Program One-Sided Violence Dataset, http://www.pcr.uu.se/research/ucdp/datasets/ucdp_one-sided_violence_dataset/. World population figures from the U.S. Census Bureau, http://www.census.gov/population/international/data/worldpop/table_population.php.

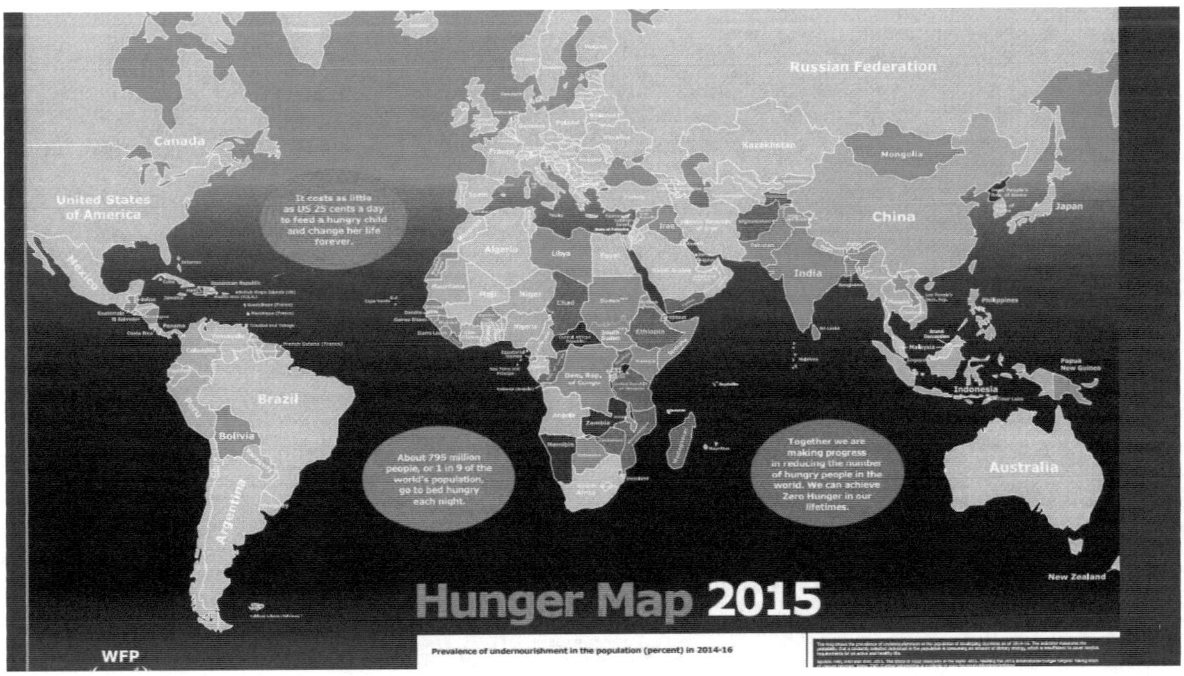
Russian Federation
Canada
United States of America
Kazakhstan
Mongolia
China
Japan
India
Algeria
Libya
Egypt
Mali
Niger
Chad
Ethiopia
Nigeria
Dem. Rep. of Congo
Zambia
Angola
Namibia
Iran
Mexico
Brazil
Bolivia
Argentina
Philippines
Indonesia
Papua New Guinea
Australia
New Zealand
It costs as little as US 25 cents a day to feed a hungry child and change her life forever.
About 795 million people, or 1 in 9 of the world's population, go to bed hungry each night.
Together we are making progress in reducing the number of hungry people in the world. We can achieve Zero Hunger in our lifetimes.
Hunger Map 2015
WFP
Prevalence of undernourishment in the population (percent) in 2014-16

END HERE

Change
Ahead

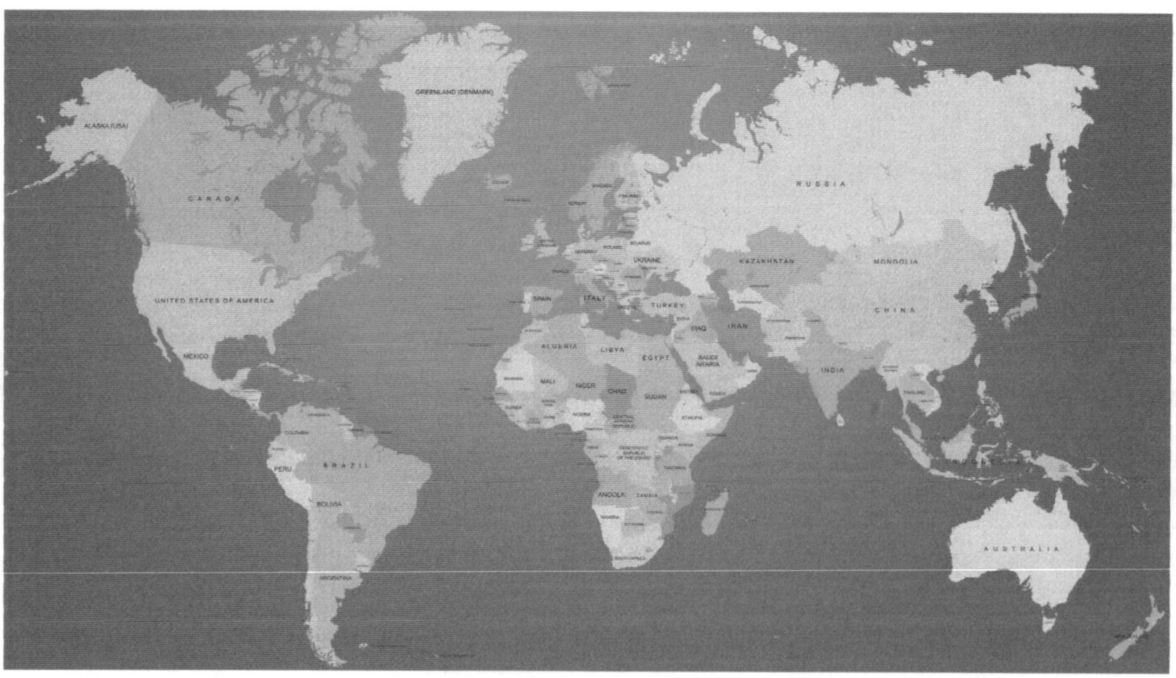
GREENLAND (DENMARK)
ALASKA (USA)
CANADA
UNITED STATES OF AMERICA
MEXICO
BRAZIL
PERU
BOLIVIA
ARGENTINA
RUSSIA
UKRAINE
KAZAKHSTAN
MONGOLIA
CHINA
SPAIN
ITALY
TURKEY
IRAQ
IRAN
ALGERIA
LIBYA
EGYPT
SAUDI ARABIA
MALI
NIGER
CHAD
SUDAN
INDIA
ANGOLA
AUSTRALIA

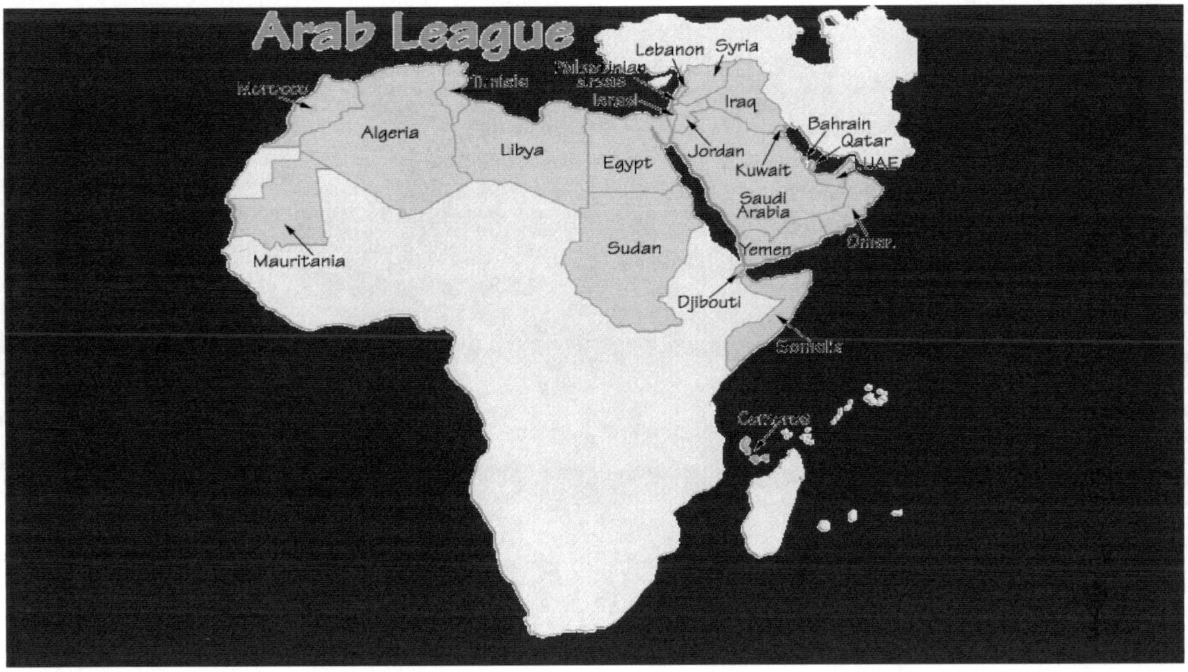
Arab League
Lebanon
Syria
Iraq
Bahrain
Qatar
UAE
Algeria
Libya
Egypt
Jordan
Kuwait
Saudi Arabia
Sudan
Yemen
Mauritania
Djibouti

With French Foreign Legion : Congo War 1977

At Ayatollah Khomeini's tomb in Iran

Travels with Secretary of State Colin Powell

With Yasser Arafat – as the world's foremost terrorist -- 1982

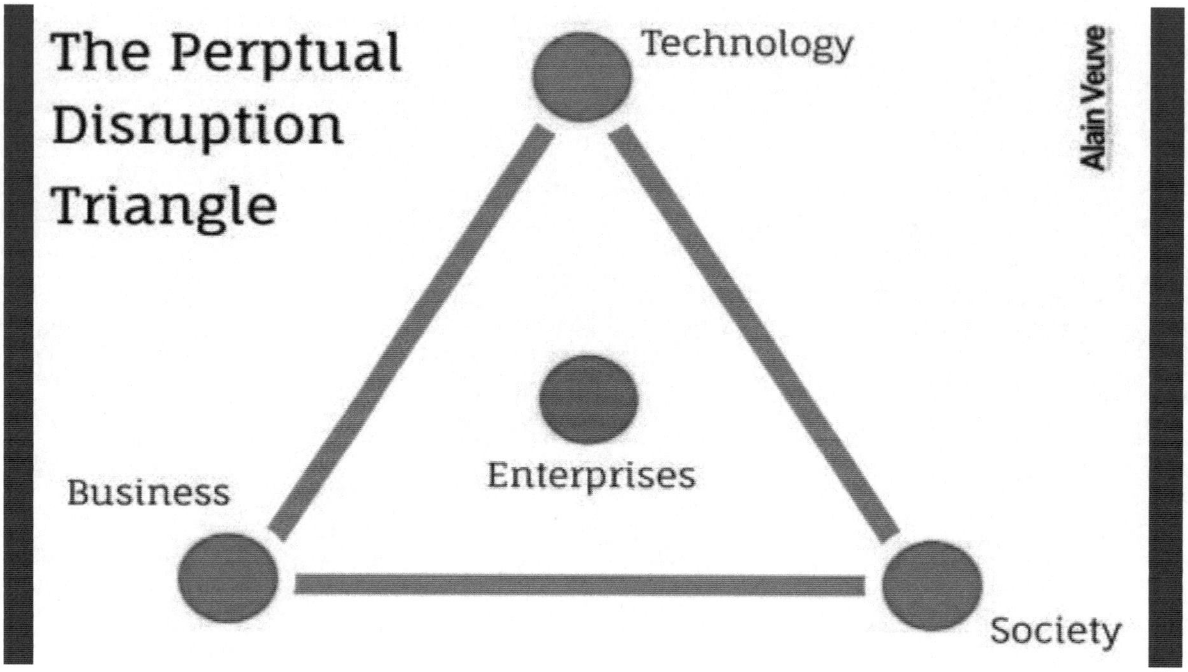
The Perptual
Disruption
Triangle
Technology
Alain Veuve
Business
Enterprises
Society

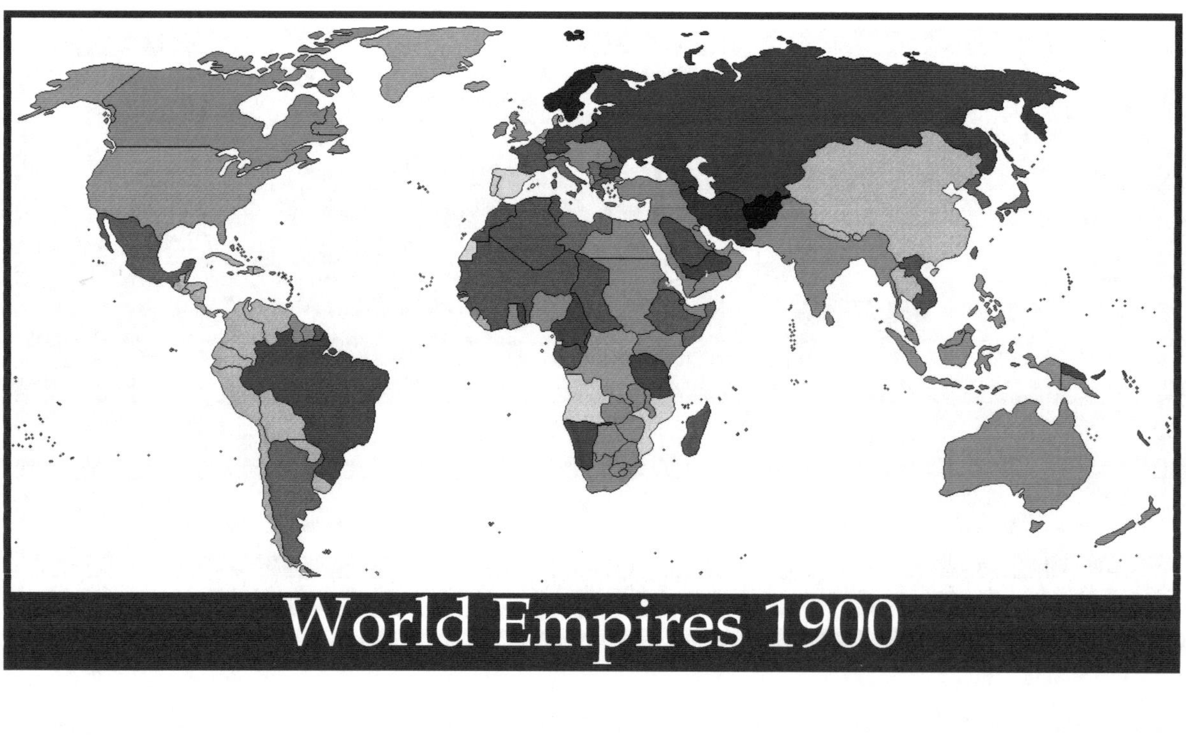
World Empires 1900

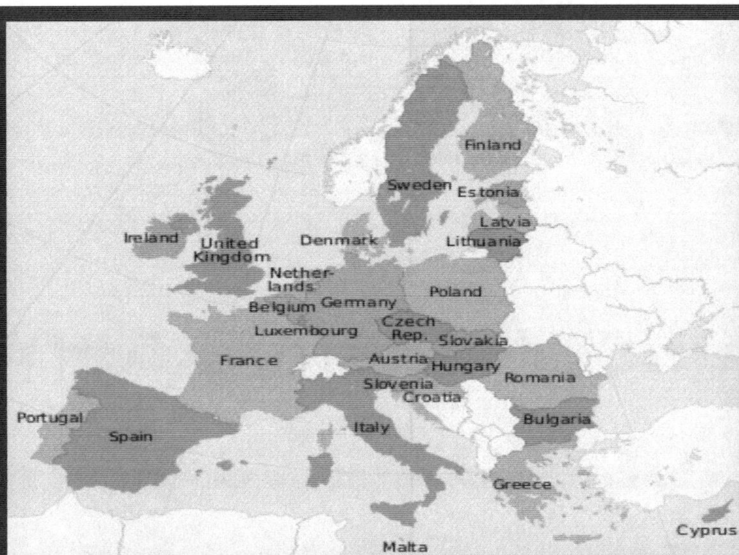

European Union : From 6 original members to 28

AZZ METAL COATINGS™

Product Advancements in Hot Dip Galvanizing

international zinc association

AGA Zn American Galvanizers Association *Protecting Steel for a Sustainable Future*

AZZ®

Corporate Profile

AZZ

Forward Looking Statements

AZZ

- Certain statements herein about our expectations of future events or results constitute forward-looking statements for purposes of the safe harbor provisions of The Private Securities Litigation Reform Act of 1995. You can identify forward-looking statements by terminology such as, "may," "should," "expects," "plans," "anticipates," "believes," "estimates," "predicts," "potential," "continue," or the negative of these terms or other comparable terminology. Such forward-looking statements are based on currently available competitive, financial and economic data and management's views and assumptions regarding future events. Such forward-looking statements are inherently uncertain, and investors must recognize that actual results may differ from those expressed or implied in the forward-looking statements. This presentation may contain forward-looking statements that involve risks and uncertainties including, but not limited to, changes in customer demand and response to products and services offered by AZZ, including demand by the power generation markets, electrical transmission and distribution markets, the industrial markets, and the hot dip galvanizing markets; prices and raw material cost, including zinc and natural gas which are used in the hot dip galvanizing process; changes in the political stability and economic conditions of the various markets that AZZ serves, foreign and domestic, customer requested delays of shipments, acquisition opportunities, currency exchange rates, adequacy of financing, and availability of experienced management and employees to implement AZZ's growth strategy. AZZ has provided additional information regarding risks associated with the business in AZZ's Annual Report on Form 10-K for the fiscal year ended February 29, 2016 and other filings with the SEC, available for viewing on AZZ's website at www.azz.com and on the SEC's website at www.sec.gov.
- You are urged to consider these factors carefully in evaluating the forward-looking statements herein and are cautioned not to place undue reliance on such forward-looking statements, which are qualified in their entirety by this cautionary statement. These statements are based on information as of the date hereof and AZZ assumes no obligation to update any forward-looking statements, whether as a result of new information, future events, or otherwise.

Corporate Overview and Vision

AZZ

- AZZ is a global provider of specialty electrical equipment and highly engineered services to the power generation, transmission, distribution and industrial markets as well as a leading provider of hot dip galvanizing services to the North American steel fabrication market
- Our Vision: We envision a world where AZZ's innovation, quality, and service minded people enable a safer and more sustainable infrastructure while providing superior returns for shareholders

Diversified Portfolio of Businesses

- Industrial Solutions (WSI and SMS Operations)
  - Specialty Welding and Overlays
  - Critical Boiler, Drum and Vessel Components
  - Boiler Construction and Overhauls
  - Structural Repair and Upgrade
  - Specialized Engineering Support
  - Corrosion/Erosion Protection

- Nuclear Solutions (NL & WSI Operations)
  - Largest supplier third-party safety related nuclear equipment (OEM Partner)
  - Equipment and Component Qualification
    - Certification – ASME III
    - Stamp – N-Certificate, NS-Certificate, NPT-Certificate
  - Alloy 600 and Stress Corrosion Cracking Mitigation
  - Engineering and Equipment Solution Development
  - Post-Fukushima

- Electrical Solutions (Legacy Electrical & Industrial Products)
  - High & Medium Voltage Bus Systems
  - Switchgear Systems
  - Enclosure Systems
  - Tubular Products
  - Hazardous and Specialty Lighting Systems

- Galvanizing
  - Largest galvanizer in North America
  - Corrosion protection services for steel fabrication industry
  - Leader in quality, dependability, and service
  - Diversified end markets

Galvanizing in North America

195 PLANTS IN US AND CANADA

- 3.7 MILLION TONNES OF STEEL PROCESSED IN 2016*
- 200,000 TONNES ZINC CONSUMED*

*Source – AGA 2016

AZZ

AZZ METAL COATINGS™

AZZ GALVANIZING

AZZ | GALVXTRA NATINA STEEL

AZZ METALIZING

AZZ | GALVABAR

AZZ | GALVXTRA COLOR

AZZ | GALVXTRA LITE

AZZ Metal Coatings Locations

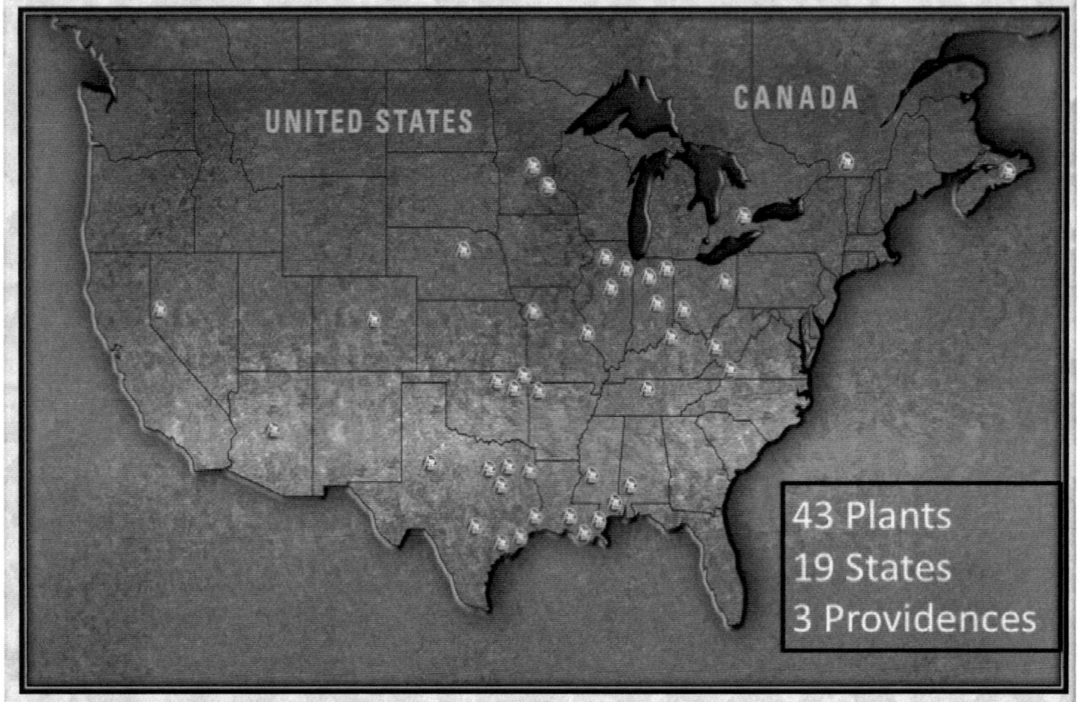

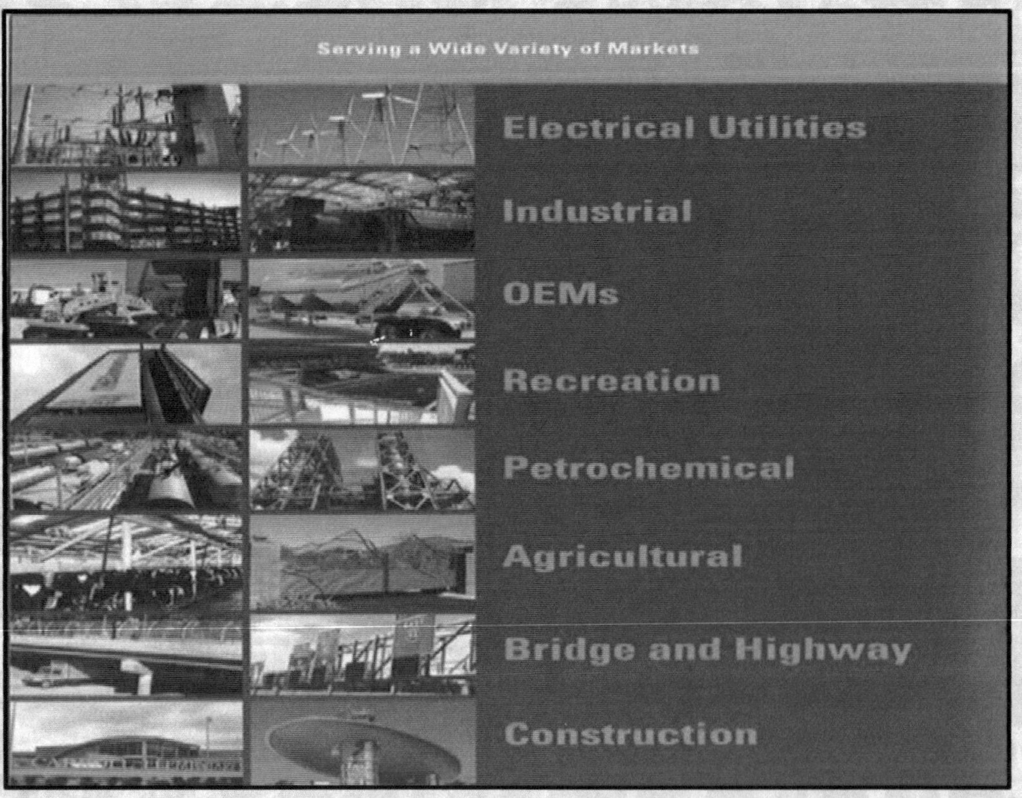

AZZ's New Products

AZZ GALVABAR — Continuous Galvanized Rebar

AZZ GALVXTRA COLOR — Powder Coating

AZZ GALVXTRA NATINA STEEL — Rustic Brown Surface Treatment

AZZ

REBAR

Epoxy coated rebar has been restricted in 4 US States and 3 Canadian provinces.

international zinc association

AZZ

Problem Solved

AZZ GALVANIZING | GALVABAR™

international zinc association

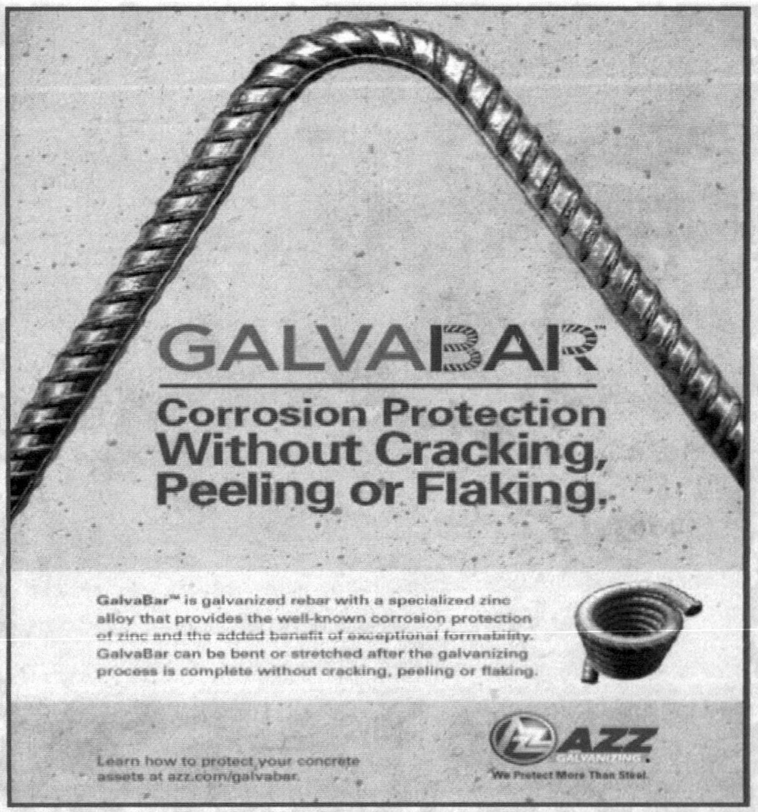

- 1st plant currently under construction.
- Will be operational early 2017.
- Long-term capital plan will support market growth.

international zinc association

Flow Coat Process w/ Modified Bath Chemistry
AZZ
GALVABAR

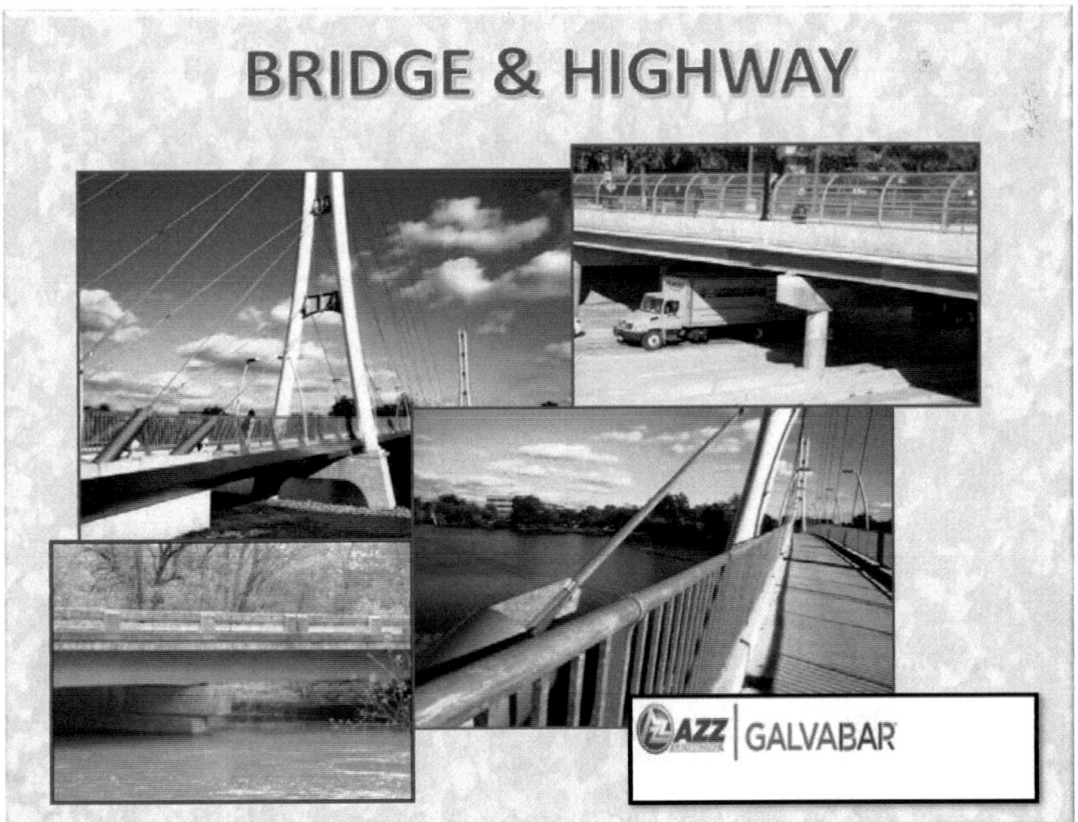
BRIDGE & HIGHWAY
AZZ
GALVABAR

AZZ | GALVABAR

AZZ GALVANIZING | GALVXTRA™

NATINA® STEEL

Architects and visual landscape professionals want the rustic brown look of weathering steel.

Problem Solved

AZZ GALVANIZING | GALVXTRA™
NATINA® STEEL

Provides look of weathering steel, but with the corrosion protection of galvanizing.

AZZ
GALVANIZING
GALVXTRA
NATINA STEEL

AZZ
GALVANIZING
GALVXTRA
NATINA STEEL

AZZ | GALVXTRA

AZZ GALVANIZING | GALVXTRA

Why Powder Coating?

- People like color.
- Attractive, durable, high- quality finish.
- Uniform coating that is compatible with galvanizing.
- Synergistic Effect with galvanizing resulting in more resistance to corrosion, chemicals and weather.
- Powder does not run or drip. No VOCs.

Thanks to the miners, the smelters, the suppliers, the brokers and the IZA for your continued support of the Galvanizing Industry
international zinc association
AGA
American Galvanizers Association
Protecting Steel for a Sustainable Future
Zn
AZZ
METAL COATINGS

Opportunities and Threats to Future Zinc Usage in the Automotive Industry

Curt D. Horvath, Technical Fellow
General Motors Company
IZA Meeting– Palm Springs, California
February 27, 2017

GM General Motors International Zinc Association

Outline

- GM's Future Materials Direction
- Competitive Threats
- Future Challenges for Zinc Coatings
- Zinc "Mega Trends"
- Opportunities for Automotive/IZA Co-Developments

GM General Motors 2 of 26 International Zinc Association

GM's Future Materials Direction

- Ultra-High Strength Steels (UHSS's)
- 3rd Gen Advanced High Strength Steels
- Aluminized Press Hardened Steels
- Mixed Material Strategy

General Motors Material Philosophy

- **Optimize Applications of Materials**
  - Right Material in the Right Application
  - Use Steel Solutions Whenever Possible
- **Mixed Material Strategy**
  - Advanced and Ultra High Strength Steels
  - Non-Ferrous Metals: Aluminum Sheet, Castings and Extrusions

GM's Focus on Efficient Fundamentals

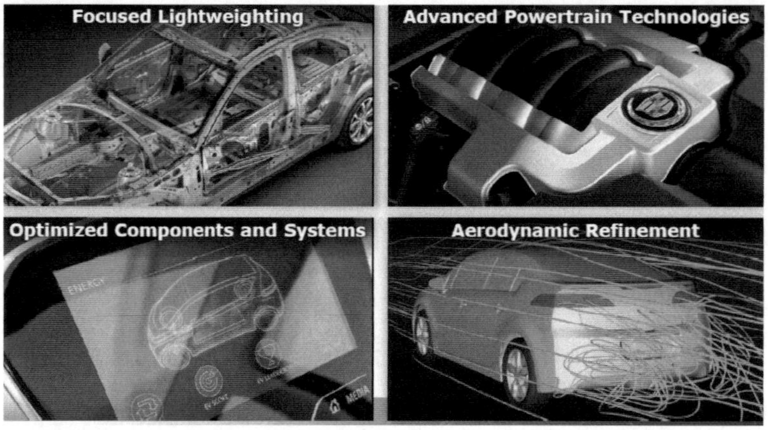

Mixed Materials – More Efficiency

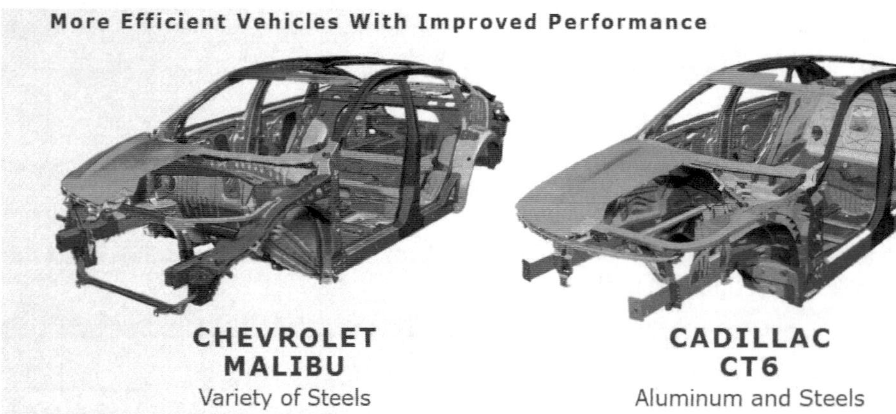

Chevy Malibu – Steel Usage

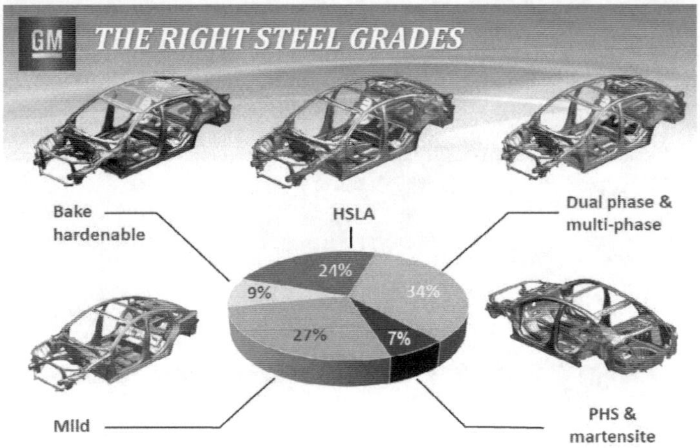

Lightweighting – Cadillac CT6

13 Different Materials

Multiple Joining Methods

Aluminum Intensive Structure

Strategic Use of High Strength and Ultra-High Strength Steels

CT6 - Strategic Use of Aluminum and Steel

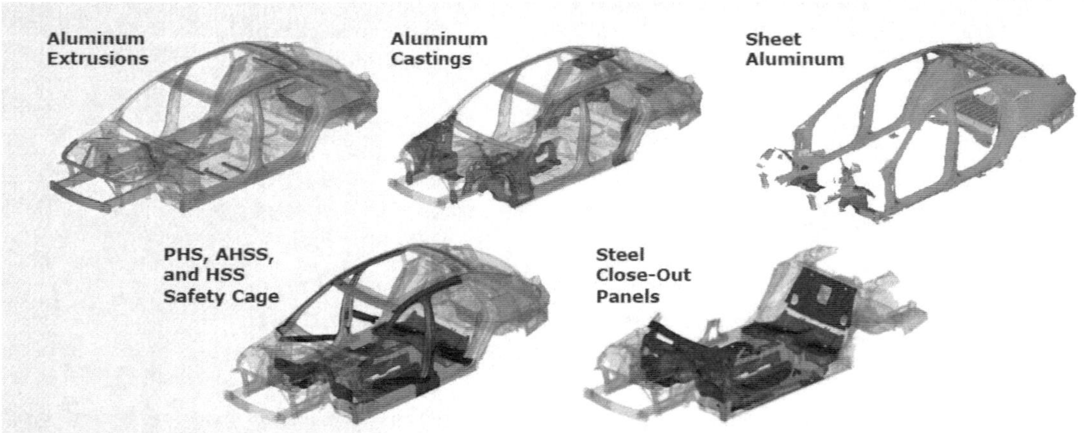

Efficient Fundamentals – Mass Reductions

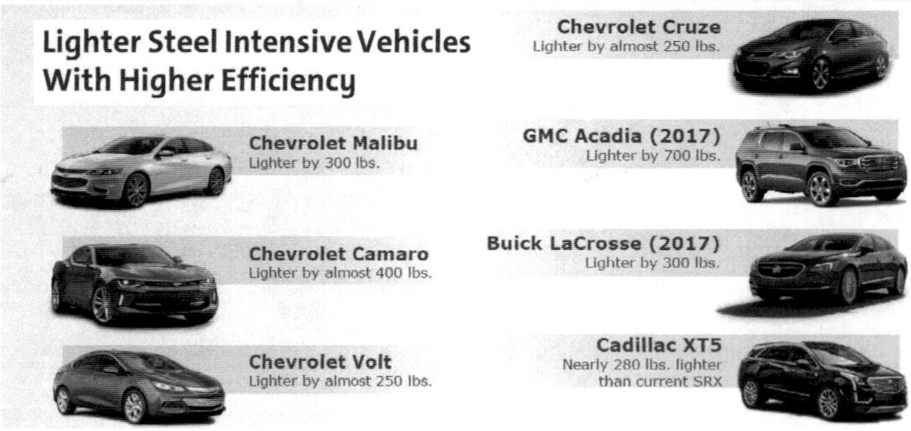

Results – Optimized Steel Vehicles

PRODUCT FOCUS

2,400 lbs. LOST

or more than 300 lbs. average per new vehicle

15 MILLION

*Gallons of fuel potentially saved per year

137k METRIC TONS

*CO_2 potentially avoided in the atmosphere per year

*internal estimates based on sales forecast and EPA estimated

3rd Gen Advanced High Strength Steels

- **GM's Direction – Fully Leverage 3rd Gen and AHSS's Advantages**
  - Active Development and Implementation of 3rd Gen AHSS's
  - Investments in infrastructure and testing/development capabilities
  - Joint research projects with industry and academia
  - Global developments
- **Challenges**
  - Hot dip coating of highly alloyed/high hardenability alloy designs
  - Spot Weld Liquid Metal Embrittlement (LME)

Competitive Threats

- AlSi Coated Press Hardened Steels (PHS)
- Aluminum and Non-Ferrous Materials
  - Al Sheet – Closures and Some Hybrid structures
  - Carbon Fiber and Advanced Composites
  - Organic Coatings
- ZnMgAl Coatings (reduced coating weights = less zinc)

Future Challenges

- Hot Dip Zinc Coated 3rd Gen Advanced Steels (HDGI)
- Laser/TIG Brazed Joints
- ZnMgAl Coatings (reduced coating weights)

Future Challenges – Hot Dip Zinc Coating of 3rd Gen Advanced Steels

- High Carbon Equivalents
  - Carbon: ~ 0.2 – 0.4%
  - Mn: ~ 2 – 3 ½ % (some grades > 10%)
- High Silicon
  - ~ 1 – 2 ½ %

→ **Difficult to Apply Hot Dip Coatings!**

Future Challenges - Welding of Zinc Coated 3rd Gen Advanced Steels

Liquid Metal Embrittlement (LME)

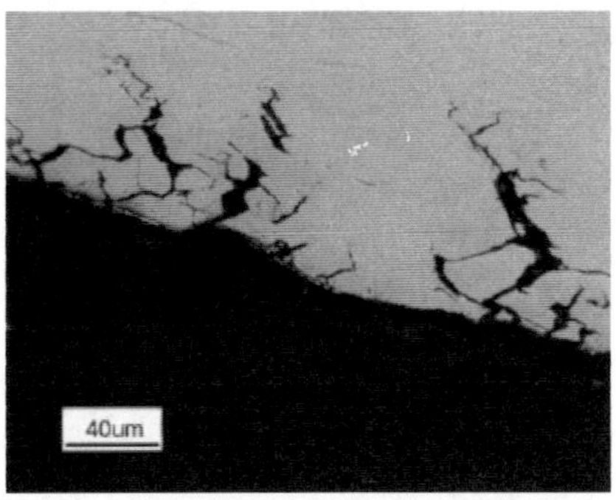

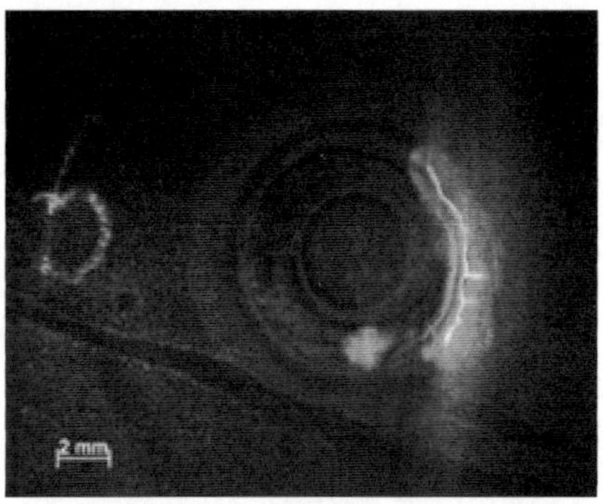

Future Challenges – Laser/TIG Brazing on HDGI

Process Overview:

- Laser beam heats filler wire (braze) above melting point
- Filler metal flows between joint by capillary action
- Issues
  - ➢ Pore Formation, Poor Wetting and Skips

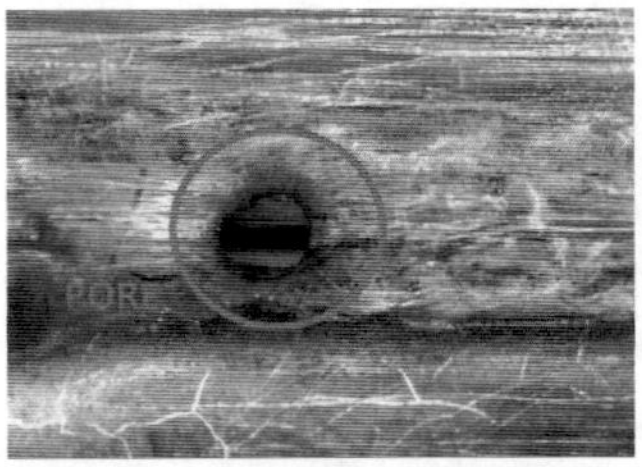

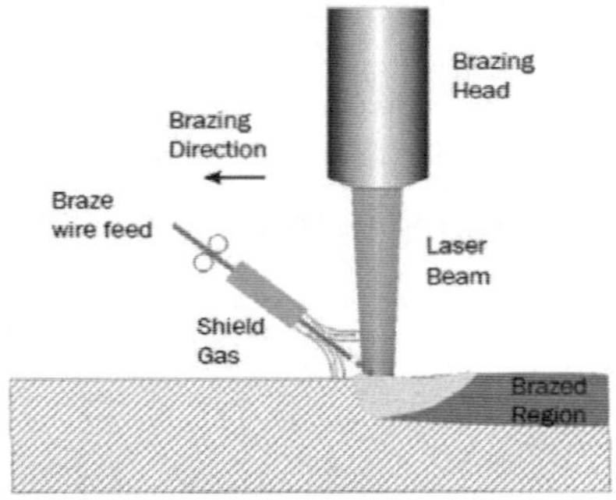

From: Fraunhofer USA, Great Designs in Steel, May 2016

Future Challenges – ZnMgAl

Significantly Reduced Coating Weight With ZnMgAl

- Improved Corrosion Protection – Especially for Al/Steel Interfaces
- Requires Additional Zinc Pot
- Coating Weight ~50% Less Than Pure Zinc
- Many Different Coating Chemistries Present Globalization Challenges

Competitive Threats - Mega Trends

Steel (zinc coatings) Face Increasing Competition......

- Hybrid Structures Will Reduce Steel/Zinc Usage
- Aluminum in Structural Applications and CFRP on Luxury and High End Vehicles, Selectively on Midsize and Smaller Vehicles.
- Aluminum on Exterior Skin Panels Likely to Increase

Competitive Threats - Mega Trends

Regulatory Environment Impact on Automotive Steel Usage

- Stable ⟹ Favors Steel Intensive Solutions
- Increasing ⟹ Favors Non-Steel Solutions

Competitive Threats - Mega Trends

- Developing Market May Not Follow Same Trends as Developed Markets
  - **Market Price Sensitivity**
  - **Availability/Infrastructure**
- Political/Geopolitical Influences – Positive and Negative!
- Carbon Footprints of Steel, Aluminum and Carbon Fiber

Future Competitive Opportunities for Steel

New Model for Steelmaking

- Reduced Emissions
  - Improved Efficiency in Steelmaking
  - Carbon Sequestration
  - Co-Generation of Electricity/Excess to Grid
- Greatly Reduced Carbon Footprint
- Carbon Sequestration/Fracking Aid in Shale Deposits

Re-Define Steel as the Green Material and Lifecycle Leader!!

Future Areas of Automotive/IZA Co-Developments

- Liquid Metal Embrittlement (LME) on 3rd Gen and High Alloy Steels
- Contact/Galvanic Corrosion (Surface Coatings)
- Laser/TIG Brazing on HDGI Coatings
- Gauge/Width Capabilities

GM General Motors 23 of 26 International Zinc Association

Recommended Actions for IZA

- Become Familiar with the Challenges Facing Zinc Coatings
- Partner with Industry and Academia
  - To Develop Welding/Joining Processes for 3rd Gen AHSS's That Address Coating Challenges and are Less Sensitive to LME!
  - To Address Surface Appearance Differences with Competitive Materials
  - To Resolve Galvanic Corrosion Issues in Mixed Metal Joints (FE: Al, Mg, CRFP)
- Be Proactive Relative to Environmental/Regulatory Issues that Affect Zinc's Future

GM General Motors 24 of 26 International Zinc Association

Conclusion

- Future of zinc coating usage in the automotive industry is clouded by many competitive threats
- A number of challenges for zinc coatings exist, such as LME on 3rd Gen steels, laser and TIG brazing issues, poor galvanic corrosion protection in hybrid structures and threats from ZnMgAl coatings
- Increasing regulations relative to fuel economy and CO2 emissions favor non-steel solutions that could have dramatic impacts on zinc usage
- Pro-active involvement of IZA in helping to solve emerging challenges will insure continued widespread use of zinc coated steels continues

C:\Users\Public\ ments\Data\ADM01

Contact Information

Curt D. Horvath
Technical Fellow
GM Technical Center, Materials and Corrosion Engineering
30200 Mound Road
Warren, MI 48092 USA
MC 480-111-3N

E-mail: curt.d.horvath@gm.com
Cell: 248-563-3394

STRATEGIC APPROACH TO ENERGY STORAGE

ViZn Energy Systems and ZAF Energy Systems

Introduction

Craig Wilkins, Founder / Executive / Chairman

- **ViZn Energy Systems** – Grid scale zinc-iron flow battery for utility and renewable applications.

- **ZAF Energy Systems**
  - Nickel Zinc battery for motive and stationary storage applications.
  - Zinc Air battery for logistics, consumer electronics, military and motive applications.

Begin with the End In Mind

ViZn ENERGY

For batteries to achieve widespread adoption:

1. **Cost** – Chemistry and materials in construction must be inexpensive and meet cost/benefit threshold of our targeted markets
2. **Safe** - Chemistry must be safe and free from toxic materials
3. **Energy Independence** - Materials must be globally abundant in quantities sufficient to meet long term needs
4. **Scale** – Cells, manufacturing and chemistry must be scalable.
5. **Robust** - Chemistry and system must have flexibility to provide wide variety of services to provide the greatest economic benefit

Electrochemical Couples – All

ViZn ENERGY

Reduction Potential / Oxidation Potential	Eo	Na+ + e- → Na	Mg+2 + 2e- → Mg	Al+3 + 3e- → Al	Ti+3 + 3e- → Ti	ZrO2 → Zr	H2O + e- → H2 +2OH-	[Zn(OH)4]-2 + 2e- → Zn	Mn+2 + 2e- → Mn	Mn+2 + 2e- → Mn	V+2 + 2e- → V	Cr+2 + 2e- → Cr	H2O + e- → O2 + 4H+ + 4e-	Zn+2 + 2e- → Zn	S + 2e- → S-2	Cr+3 + e- → Cr+2	Pb+2 + 2e- → Pb	Co+2 + 2e- → Co	Ni+2 + 2e- → Ni	V+3 + e- → V+2	Mo+3 + 3e- → Mo	Sn+2 + 2e- → Sn	2H+ + 2e- → H2	Cu+2 + 2e- → Cu	$[Fe(CN)_6]^{-3}$ + e- → $[Fe(CN)_6]^{-4}$	Fe^{+3} + e^- → Fe^{+2}	VO_2^{+2} + $2e^-$ → VO^{+2}	Br_2 + 2e- → $2Br^{-1}$	Cl2 + 2e- → 2Cl-	Cr2O7-2 + 4e- → Cr+3	MnO4- + 5e- → Mn+2	NiO+2 + 2e- → Ni+2	PbO2 + 2e- → PbSO4	Co+3 + e- → Co+2
		-2.71	-2.36	-1.67	-1.63	-1.55	-1.53	-1.28	-1.18	-1.18	-1.13	-0.90	-0.83	-0.76	-0.47	-0.42	-0.36	-0.28	-0.26	-0.26	-0.20	-0.16	0.00	0.34	0.36	0.70	1.00	1.07	1.36	1.38	1.51	1.59	1.69	1.92
Li → Li+ + e-	3.05	0.33	0.69	1.38	1.42	1.49	1.52	1.77	1.87	1.87	1.92	2.15	2.22	2.29	2.58	2.62	2.69	2.77	2.79	2.79	2.85	2.89	3.05	3.39	3.41	3.75	4.05	4.12	4.41	4.43	4.56	4.64	4.74	4.97
Ca → Ca+2 + 2e-	2.84	0.13	0.48	1.17	1.21	1.29	1.31	1.56	1.66	1.66	1.71	1.94	2.01	2.08	2.37	2.42	2.48	2.56	2.58	2.58	2.64	2.68	2.84	3.18	3.20	3.54	3.84	3.91	4.20	4.22	4.35	4.43	4.53	4.76
Na → Na^+ +e-	2.71	[illegible]	0.36	1.04	1.08	1.16	1.18	1.43	1.53	1.53	1.58	1.81	1.88	1.95	2.24	2.29	2.36	2.44	2.46	2.45	2.51	2.55	2.71	3.05	3.07	3.41	3.71	3.78	4.07	4.09	4.22	4.31	4.40	4.63
Mg → Mg^{+2} +2e-	2.36	[illegible]	[illegible]	0.69	0.73	0.80	0.83	1.08	1.18	1.18	1.23	1.46	1.53	1.60	1.89	1.93	2.00	2.08	2.10	2.10	2.16	2.20	2.36	2.70	2.72	3.06	3.36	3.43	3.72	3.74	3.87	3.95	4.05	4.28
Al → Al^{+3} + 3e-	1.67	[illegible]	[illegible]	[illegible]	0.04	0.12	0.14	0.39	0.49	0.49	0.54	0.77	0.84	0.91	1.20	1.25	1.31	1.39	1.41	1.41	1.47	1.51	1.67	2.01	2.03	2.37	2.67	2.74	3.03	3.05	3.18	3.26	3.36	3.59
Ti → Ti+3 + 3e-	1.63	[illegible]	[illegible]	[illegible]	[illegible]	0.08	0.10	0.35	0.45	0.45	0.50	0.73	0.80	0.87	1.16	1.21	1.27	1.35	1.37	1.37	1.43	1.47	1.63	1.97	1.99	2.33	2.63	2.70	2.99	3.01	3.14	3.22	3.32	3.55
Zr → Zr+4 + 4e-	1.55	[illegible]	[illegible]	[illegible]	[illegible]	[illegible]	[illegible]	0.27	0.37	0.37	0.42	0.65	0.72	0.79	1.08	1.13	1.19	1.28	1.30	1.29	1.35	1.39	1.55	1.89	1.91	2.25	2.55	2.62	2.91	2.93	3.06	3.15	3.24	3.47
Zn → [Zn(OH)4]-2 + 2e-	1.28	[illegible]	[illegible]	[illegible]	[illegible]	[illegible]	[illegible]	[illegible]	0.10	0.10	0.15	0.38	0.45	0.52	0.81	0.86	0.92	1.00	1.02	1.02	1.08	1.12	1.28	1.62	1.64	1.98	2.28	2.35	2.64	2.66	2.79	2.87	2.97	3.20
Mn → Mn+2 +2e-	1.18	[illegible]	[illegible]	[illegible]	[illegible]	[illegible]	[illegible]	[illegible]	[illegible]	[illegible]	[illegible]	0.28	0.35	0.42	0.71	0.76	0.82	0.90	0.92	0.92	0.98	1.02	1.18	1.52	1.54	1.88	2.18	2.25	2.54	2.56	2.69	2.77	2.87	3.10
Cr → Cr+2 +2e-	0.90	[illegible]	[illegible]	[illegible]	[illegible]	[illegible]	[illegible]	[illegible]	[illegible]	[illegible]	[illegible]	[illegible]	0.07	0.14	0.43	0.48	0.54	0.62	0.64	0.64	0.70	0.74	0.90	1.24	1.26	1.60	1.90	1.97	2.26	2.28	2.41	2.49	2.59	2.82
Zn → Zn+2 + 2e-	0.76	[illegible]	[illegible]	[illegible]	[illegible]	[illegible]	[illegible]	[illegible]	[illegible]	[illegible]	[illegible]	[illegible]	[illegible]	[illegible]	0.29	0.34	0.40	0.48	0.50	0.50	0.56	0.60	0.76	1.10	1.12	1.46	1.76	1.83	2.12	2.14	2.27	2.35	2.45	2.68
S-2 → S + 2e-	0.47	[illegible]	[illegible]	[illegible]	[illegible]	[illegible]	[illegible]	[illegible]	[illegible]	[illegible]	[illegible]	[illegible]	[illegible]	[illegible]	[illegible]	0.05	0.11	0.19	0.21	0.21	0.27	0.31	0.47	0.81	0.83	1.17	1.47	1.54	1.83	1.85	1.98	2.06	2.16	2.39
Cr+2 → Cr+3 +e-	0.42	[illegible]	[illegible]	[illegible]	[illegible]	[illegible]	[illegible]	[illegible]	[illegible]	[illegible]	[illegible]	[illegible]	[illegible]	[illegible]	[illegible]	[illegible]	0.07	0.15	0.17	0.16	0.22	0.26	0.42	0.76	0.78	1.12	1.42	1.49	1.78	1.80	1.93	2.02	2.11	2.34
Pb → Pb+2 +2e-	0.36	[illegible]	[illegible]	[illegible]	[illegible]	[illegible]	[illegible]	[illegible]	[illegible]	[illegible]	[illegible]	[illegible]	[illegible]	[illegible]	[illegible]	[illegible]	[illegible]	[illegible]	0.10	0.10	0.16	0.20	0.36	0.70	0.72	1.06	1.36	1.43	1.72	1.74	1.87	1.95	2.05	2.28
Co → Co+2 +2e-	0.28	[illegible]	[illegible]	[illegible]	[illegible]	[illegible]	[illegible]	[illegible]	[illegible]	[illegible]	[illegible]	[illegible]	[illegible]	[illegible]	[illegible]	[illegible]	[illegible]	[illegible]	[illegible]	[illegible]	[illegible]	0.12	0.28	0.62	0.64	0.98	1.28	1.35	1.64	1.66	1.79	1.87	1.97	2.20
Ni → Ni+2 + 2e-	0.26	[illegible]	[illegible]	[illegible]	[illegible]	[illegible]	[illegible]	[illegible]	[illegible]	[illegible]	[illegible]	[illegible]	[illegible]	[illegible]	[illegible]	[illegible]	[illegible]	[illegible]	[illegible]	[illegible]	[illegible]	0.10	0.26	0.60	0.62	0.96	1.26	1.33	1.62	1.64	1.77	1.85	1.95	2.18
V+2 → V+3 +e-	0.26	[illegible]	[illegible]	[illegible]	[illegible]	[illegible]	[illegible]	[illegible]	[illegible]	[illegible]	[illegible]	[illegible]	[illegible]	[illegible]	[illegible]	[illegible]	[illegible]	[illegible]	[illegible]	[illegible]	[illegible]	0.10	0.26	0.60	0.62	0.96	1.26	1.33	1.62	1.64	1.77	1.85	1.95	2.18
Mo → Mo+3 +3e-	0.20	[illegible]	[illegible]	[illegible]	[illegible]	[illegible]	[illegible]	[illegible]	[illegible]	[illegible]	[illegible]	[illegible]	[illegible]	[illegible]	[illegible]	[illegible]	[illegible]	[illegible]	[illegible]	[illegible]	[illegible]	[illegible]	0.20	0.54	0.56	0.90	1.20	1.27	1.56	1.58	1.71	1.79	1.89	2.12
Sn → Sn+2 +2e-	0.16	[illegible]	[illegible]	[illegible]	[illegible]	[illegible]	[illegible]	[illegible]	[illegible]	[illegible]	[illegible]	[illegible]	[illegible]	[illegible]	[illegible]	[illegible]	[illegible]	[illegible]	[illegible]	[illegible]	[illegible]	[illegible]	0.16	0.50	0.52	0.86	1.16	1.23	1.52	1.54	1.67	1.75	1.85	2.08
Cu → Cu+2 + 2e-	-0.34	[illegible]	[illegible]	[illegible]	[illegible]	[illegible]	[illegible]	[illegible]	[illegible]	[illegible]	[illegible]	[illegible]	[illegible]	[illegible]	[illegible]	[illegible]	[illegible]	[illegible]	[illegible]	[illegible]	[illegible]	[illegible]	[illegible]	[illegible]	[illegible]	0.36	0.66	0.73	1.02	1.04	1.17	1.25	1.35	1.58
$[Fe(CN)_6]^{-3}$ → $[Fe(CN)6]^{-2}$ + e-	-0.36	[illegible]	[illegible]	[illegible]	[illegible]	[illegible]	[illegible]	[illegible]	[illegible]	[illegible]	[illegible]	[illegible]	[illegible]	[illegible]	[illegible]	[illegible]	[illegible]	[illegible]	[illegible]	[illegible]	[illegible]	[illegible]	[illegible]	[illegible]	[illegible]	0.34	0.64	0.71	1.00	1.02	1.15	1.23	1.33	1.56
OH^- → O_2 + 4e- +$2H_2O$	-0.40	[illegible]	[illegible]	[illegible]	[illegible]	[illegible]	[illegible]	[illegible]	[illegible]	[illegible]	[illegible]	[illegible]	[illegible]	[illegible]	[illegible]	[illegible]	[illegible]	[illegible]	[illegible]	[illegible]	[illegible]	[illegible]	[illegible]	[illegible]	[illegible]	0.30	0.60	0.67	0.96	0.98	1.11	1.19	1.29	1.52
Fe^{+2} → Fe^{+3} + e-	-0.70	[illegible]	[illegible]	[illegible]	[illegible]	[illegible]	[illegible]	[illegible]	[illegible]	[illegible]	[illegible]	[illegible]	[illegible]	[illegible]	[illegible]	[illegible]	[illegible]	[illegible]	[illegible]	[illegible]	[illegible]	[illegible]	[illegible]	[illegible]	[illegible]	[illegible]	0.30	0.37	0.66	0.68	0.81	0.89	0.99	1.22
VO^{+2} $\|VO^{2+2}$ + e-	-1.00	[illegible]	[illegible]	[illegible]	[illegible]	[illegible]	[illegible]	[illegible]	[illegible]	[illegible]	[illegible]	[illegible]	[illegible]	[illegible]	[illegible]	[illegible]	[illegible]	[illegible]	[illegible]	[illegible]	[illegible]	[illegible]	[illegible]	[illegible]	[illegible]	[illegible]	[illegible]	0.07	0.36	0.38	0.51	0.59	0.69	0.92
OH - → O2 + 4e- +2H2O	-1.00	[illegible]	[illegible]	[illegible]	[illegible]	[illegible]	[illegible]	[illegible]	[illegible]	[illegible]	[illegible]	[illegible]	[illegible]	[illegible]	[illegible]	[illegible]	[illegible]	[illegible]	[illegible]	[illegible]	[illegible]	[illegible]	[illegible]	[illegible]	[illegible]	[illegible]	[illegible]	0.07	0.36	0.38	0.51	0.59	0.69	0.92
2Br- → Br2 + 2e-	-1.07	[illegible]	[illegible]	[illegible]	[illegible]	[illegible]	[illegible]	[illegible]	[illegible]	[illegible]	[illegible]	[illegible]	[illegible]	[illegible]	[illegible]	[illegible]	[illegible]	[illegible]	[illegible]	[illegible]	[illegible]	[illegible]	[illegible]	[illegible]	[illegible]	[illegible]	[illegible]	[illegible]	0.29	0.31	0.44	0.52	0.62	0.85
H2O → O2 + 4H+ + 4e-	-1.23	[illegible]	[illegible]	[illegible]	[illegible]	[illegible]	[illegible]	[illegible]	[illegible]	[illegible]	[illegible]	[illegible]	[illegible]	[illegible]	[illegible]	[illegible]	[illegible]	[illegible]	[illegible]	[illegible]	[illegible]	[illegible]	[illegible]	[illegible]	[illegible]	[illegible]	[illegible]	[illegible]	0.13	0.15	0.28	0.36	0.46	0.69
PbSO4 → PbO2 + 2e-	-1.69	[illegible]	[illegible]	[illegible]	[illegible]	[illegible]	[illegible]	[illegible]	[illegible]	[illegible]	[illegible]	[illegible]	[illegible]	[illegible]	[illegible]	[illegible]	[illegible]	[illegible]	[illegible]	[illegible]	[illegible]	[illegible]	[illegible]	[illegible]	[illegible]	[illegible]	[illegible]	[illegible]	[illegible]	[illegible]	[illegible]	[illegible]	[illegible]	0.23
H2O → O2 + 4H+ overpotential	-1.80	[illegible]	[illegible]	[illegible]	[illegible]	[illegible]	[illegible]	[illegible]	[illegible]	[illegible]	[illegible]	[illegible]	[illegible]	[illegible]	[illegible]	[illegible]	[illegible]	[illegible]	[illegible]	[illegible]	[illegible]	[illegible]	[illegible]	[illegible]	[illegible]	[illegible]	[illegible]	[illegible]	[illegible]	[illegible]	[illegible]	[illegible]	[illegible]	0.12

Electrochemical Couples – Aqueous (flow battery)

ViZn ENERGY

	Co+2 + 2e- → Co	Ni+2 + 2e- → Ni	V+3 + e- → V+2	Mo+3 + 3e- → Mo	Sn+2 + 2e- → Sn	2H+ + 2e- → H2	Cu+2 + 2e- → Cu	$[Fe(CN)_6]^{-3}$ + e- → $[Fe(CN)_6]^{-4}$	Fe^{+3} + e^- → Fe^{+2}	VO_2^{+2} + $2e^-$ → VO^{+2}	Br_2 + 2e- → $2Br^{-1}$	Cl2 + 2e- → 2Cl-	PbO2 + 2e- → PbSO4
	-0.28	-0.26	-0.26	-0.20	-0.16	0.00	0.34	0.36	0.70	1.00	1.07	1.36	1.69
Zn → [Zn(OH)4]-2 + 2e-	1.00	1.02	1.02	1.08	1.12	1.28	1.62	**1.64**	1.98	2.28	2.35	2.64	2.97
Mn → Mn+2 +2e-	0.90	0.92	0.92	0.98	1.02	1.18	1.52	1.54	1.88	2.18	2.25	2.54	2.87
Cr → Cr+2 +2e-	0.62	0.64	0.64	0.70	0.74	0.90	1.24	1.26	1.60	1.90	1.97	2.26	2.59
Zn → Zn+2 + 2e-	0.48	0.50	0.50	0.56	0.60	0.76	1.10	1.12	1.46	1.76	**1.83**	2.12	2.45
S-2 → S + 2e-	0.19	0.21	0.21	0.27	0.31	0.47	0.81	0.83	1.17	1.47	1.54	1.83	2.16
Cr+2 → Cr+3 +e-	0.15	0.17	0.16	0.22	0.26	0.42	0.76	0.78	1.12	1.42	1.49	1.78	2.11
Pb → Pb+2 +2e-	0.08	0.10	0.10	0.16	0.20	0.36	0.70	0.72	1.06	1.36	1.43	1.72	**2.05**
Co → Co+2 +2e-	[illegible]	[illegible]	[illegible]	0.08	0.12	0.28	0.62	0.64	0.98	1.28	1.35	1.64	1.97
Ni → Ni+2 + 2e-	[illegible]	[illegible]	[illegible]	0.06	0.10	0.26	0.60	0.62	0.96	1.26	1.33	1.62	1.95
V+2 → V+3 +e-	[illegible]	[illegible]	[illegible]	0.06	0.10	0.26	0.60	0.62	0.96	**1.26**	1.33	1.62	1.95
Mo → Mo+3 +3e-	[illegible]	[illegible]	[illegible]	[illegible]	0.04	0.20	0.54	0.56	0.90	1.20	1.27	1.56	1.89
Sn → Sn+2 +2e-	[illegible]	[illegible]	[illegible]	[illegible]	[illegible]	0.16	0.50	0.52	0.86	1.16	1.23	1.52	1.85
Cu → Cu+2 + 2e-	[illegible]	[illegible]	[illegible]	[illegible]	[illegible]	[illegible]	[illegible]	[illegible]	0.36	0.66	0.73	1.02	1.35
$[Fe(CN)_6]^{-3}$ → $[Fe(CN)6]^{-2}$ + e^-	[illegible]	[illegible]	[illegible]	[illegible]	[illegible]	[illegible]	[illegible]	[illegible]	0.34	0.64	0.71	1.00	1.33
OH^- → O_2 + 4e- +$2H_2O$	[illegible]	[illegible]	[illegible]	-0.60	[illegible]	[illegible]	[illegible]	[illegible]	0.30	0.60	0.67	0.96	1.29
Fe^{+2} → Fe^{+3} + e^-	-0.98	-0.96	-0.96	-0.90	-0.86	-0.70	-0.36	-0.34	0.00	0.30	0.37	0.66	0.99

Flow Battery Benefits:

- Decouple power/energy
- Superior thermal mgmt
- De-energize quickly
- Long life
- Life extension w/new stacks

Electrochemical Couples – Abundance

ViZn ENERGY

	% of 2035 storage demand		Cu^{+2} + $2e^-$ → Cu	$[Fe(CN)_6]^{-3}$ + e- → $[Fe(CN)_6]^{-4}$	Fe^{+3} + e^- → Fe^{+2}	VO_2^{+2} + $2e^-$ → VO^{+2}	Br_2 + 2e- → $2Br^{-1}$	NiO+2 + 2e- → Ni+2	PbO_2 + $2e^-$ → $PbSO_4$
	US	world	0.34	0.36	0.70	1.00	1.07	1.59	1.69
Zn → $[Zn(OH)_4]^{-2}$ + $2e^-$	102%	218%	1.62	1.64	1.98	2.28	2.35	2.87	2.97
Zn → Zn^{+2} + $2e^-$	111%	237%	1.10	1.12	1.46	1.76	1.83	2.35	2.45
S^{-2} → S + $2e^-$			0.81	0.83	1.17	1.47	1.54	2.06	2.16
Cr^{+2} → Cr^{+3} + e^-	5%	427%	0.76	0.78	1.12	1.42	1.49	2.02	2.11
Pb → Pb^{+2} + $2e^-$	15%	23%	0.70	0.72	1.06	1.36	1.43	1.95	**2.05**
Ni → Ni^{+2} + $2e^-$	0%	68%	0.60	0.62	0.96	1.26	1.33	**1.85**	1.95
V^{+2} → V^{+3} + e^-	0%	7%	0.60	0.62	0.96	1.26	1.33	1.85	1.95
Cu → Cu^{+2} + $2e^-$	114%	263%	0.00	0.02	0.36	0.66	0.73	1.25	1.35

Zn/Fe Redox Benefits:

- Abundant chemistry
- Safe chemistry
- Low cost chemistry
- Good electrochemical couple
- Alkaline chemistry lowers material cost
- Alkaline chemistry reduces dendrite problems

Electrochemical Couples – Abundance

ViZn ENERGY

battery metal	commercial reserves (in millions of tonnes)		cost ($/kg)	% of 2035 storage demand (no carbon fuels, all RES)		cost of metal per trillion Wh (B$)
	US	world		US	world	
lithium	0.04	10.1	85	4%	104%	6.3
nickel	0	71	13	0%	102%	9.0
vanadium	0.05	13	48	0%	7%	61.9
lead	7.7	79	2	24%	29%	4.3
zinc	14	230	2	127%	217%	1.5

Electrochemical Couples – Abundance

ViZn ENERGY

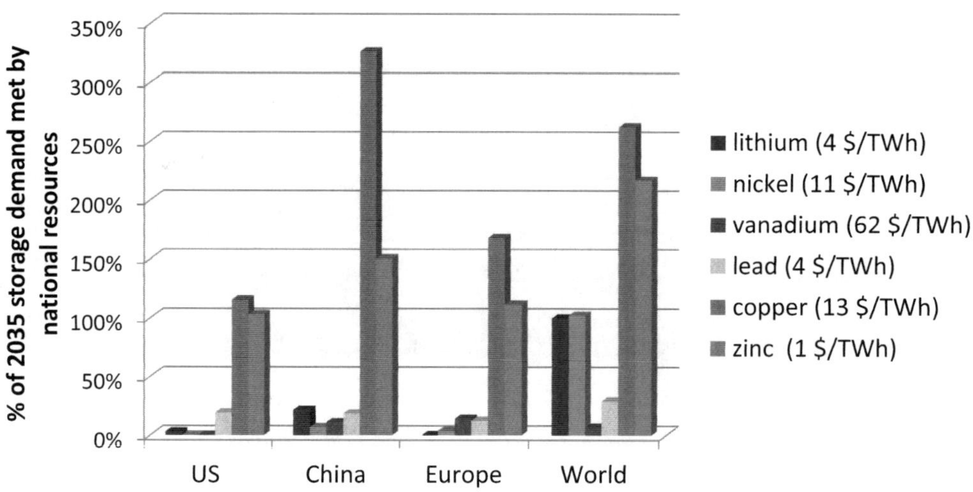

Zinc Consumption For Storage

	ViZn ENERGY	ZAF Energy Systems
Chemistry	Zinc/Iron	Nickel/Zinc
Hours of Capacity	3	2
Pounds of Zinc/Battery	1,480	15
Batteries per MW	5	1,088
Zinc per Megawatt	7,400 lbs.	16,320 lbs.
Zinc per Gigawatt	7,400,000 lbs.	16,320,000 lbs.
Metric Tons / GW	3,363	7,418
Metric Tons / TW	3,363,000	7,418,000

Zinc Production and Reserves (2014)

World Mine Production and Reserves: Reserves estimates for Bolivia, Canada, Ireland, Mexico, and other countries were revised based on company data. The reserves estimates for Australia and Peru were revised based on data from Government reports.

	Mine production[7]		Reserves[8]
	2013	2014[e]	
United States	784	820	10,000
Australia	1,520	1,500	[9]62,000
Bolivia	407	410	4,500
Canada	426	350	5,900
China	5,000	5,000	43,000
India	793	700	11,000
Ireland	327	300	1,100
Kazakhstan	362	330	10,000
Mexico	643	700	16,000
Peru	1,350	1,300	29,000
Other countries	1,800	1,900	42,000
World total (rounded)	13,400	13,300	230,000

*** Millions of Metric Tons**

SUSTAINABILITY FACTORS

Environmental, Material Safety, Recycling

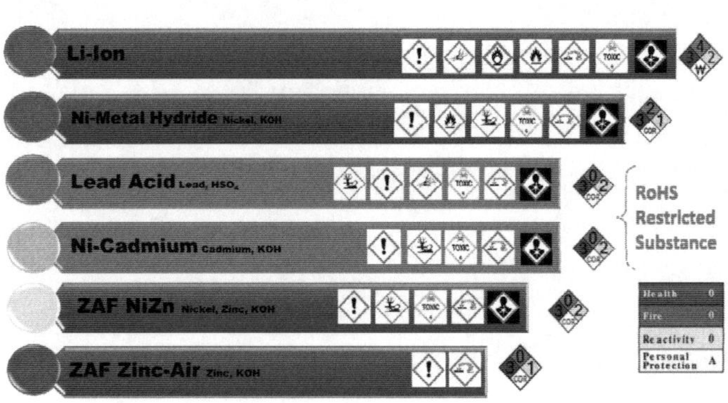

85% Material Recovery yields
40% Reduction in Battery Cost

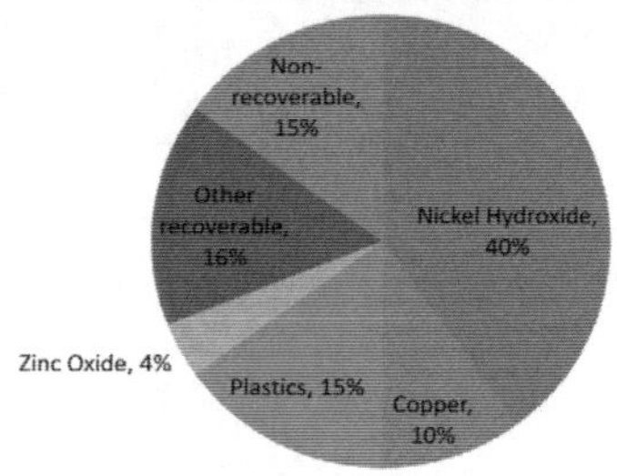

Report Exposes Lead Hazards in Developing World

Man standing on a pile of used lead-acid batteries. Some experts believe that lead poisoning from the improper recycling of used lead-acid batteries is the #1 childhood environmental health threat globally.

A man in Tegal, Indonesia, uses a hammer to break up old batteries by hand in order to extract the valuable lead within. The improper recycling of used lead-acid batteries takes place in almost every low- and middle-income country. Toxic lead is spilled everywhere in the process.

- Battery recycling contributes to more than 150 sites in the Pure Earth database, putting almost 1 million people at risk related directly to lead, mercury or cadmium pollutants.
- Geographically, the largest numbers of polluted sites are in Southeast Asia, with Africa, Central and South America also contributing a substantial amount.

Source: "2016 World's Worst Pollution Problems"

Summary

- **Market** – Energy storage is rapidly growing and provides a unique strategic opportunity for growth.
- **Innovation** – Energy storage is the enabling technology to meet the world's demand for sustainable energy.
- **Collaboration –** Strategic alignment between mining communities and technology developers will accelerate the rate of innovation and market growth.

Thank you

Craig Wilkins
Phone: (509) 939-9309
Email: craig.wilkins@viznenergy.com or craig.wilkins@zafsys.com

WELLS FARGO SECURITIES

What a Difference a Year Makes...

Implications for Metals/Mining Post a Credit Market Recovery

Senior Analyst: David Deterding
(704) 410-3234 / david.deterding@wellsfargo.com

Senior Analyst: Tyler Gately
(704) 410-3369 / tyler.gately@wellsfargo.com

February 2017

Please see page 16 for the rating definitions, important disclosures and required analyst certifications.
This report is available on wellsfargoresearch.com and on Bloomberg WFRE.
All estimates/forecasts are as of 2/27/17 unless otherwise stated.

Macro Backdrop

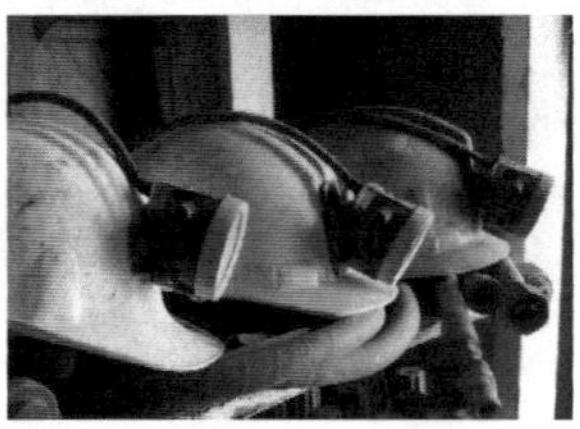

Global Growth Prospects Return After Tumultuous 2016

Global PMIs Back Above 50

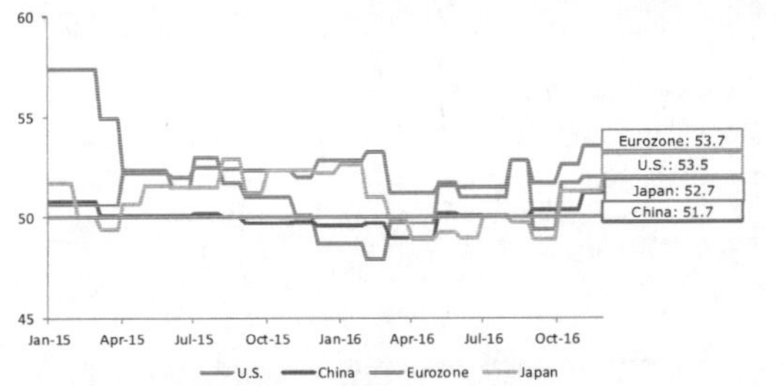

Source: Bloomberg, Wells Fargo Securities.

Global GDPs Revisions Move Higher

		2013	2014	2015	2016	2017E 12/31/15	2017E 6/30/16	2017E Current	2018E
OECD	United States	1.9%	2.4%	2.4%	1.6%	2.1%	1.8%	2.3%	2.5%
	EuroZone	0.4%	0.9%	1.5%	1.6%	2.6%	2.5%	1.5%	1.5%
	Germany	0.4%	1.4%	1.7%	1.9%	1.7%	1.4%	1.6%	1.6%
	Italy	-1.9%	-0.3%	0.7%	0.9%	1.2%	1.0%	0.7%	0.1%
	France	0.2%	0.4%	1.1%	1.2%	1.5%	1.2%	1.3%	1.4%
	Japan	1.5%	-0.1%	1.2%	1.0%	0.3%	0.1%	0.8%	0.9%
BRIC	Brazil	2.5%	0.1%	-3.7%	-3.5%	0.0%	0.5%	0.2%	2.2%
	Russia	1.3%	0.6%	-4.0%	-0.6%	1.0%	1.0%	1.2%	1.5%
	India	5.0%	6.9%	7.5%	7.6%	7.5%	7.4%	7.2%	7.7%
	China	7.7%	7.4%	6.9%	6.7%	6.0%	6.2%	6.5%	6.0%
World		3.0%	3.4%	2.9%	3.1%	3.6%	3.4%	3.4%	3.6%

Source: World Bank, Bloomberg, Wells Fargo Securities.

Commodities Rallying Off Lows as the USD Holds Near Highs

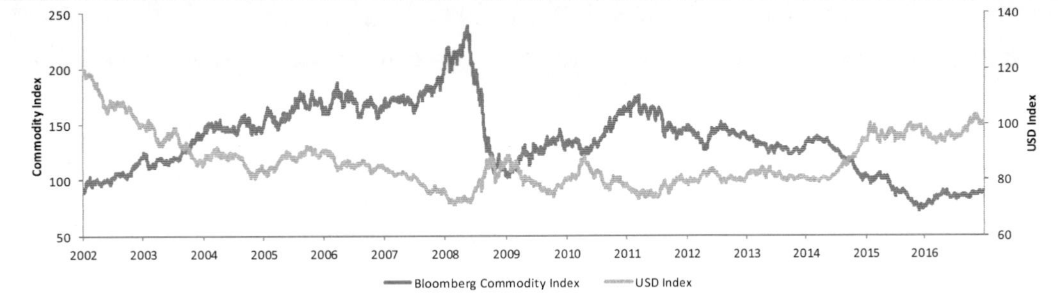

Source: Bloomberg, Wells Fargo Securities.

Source: Wells Fargo Securities and Bloomberg L.P.

The Chinese Economy Continues its Managed Slowdown

GDP Growth Has Been Gradually Slowing Since 2008

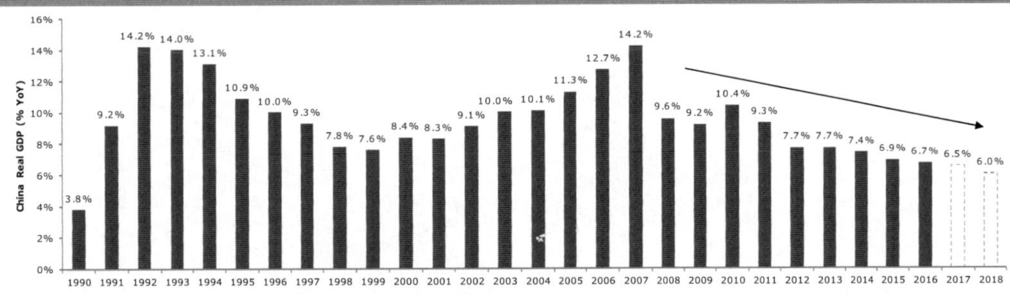

Source: Bloomberg, Wells Fargo Securities.

But China Remains the Top Consumer of Commodities

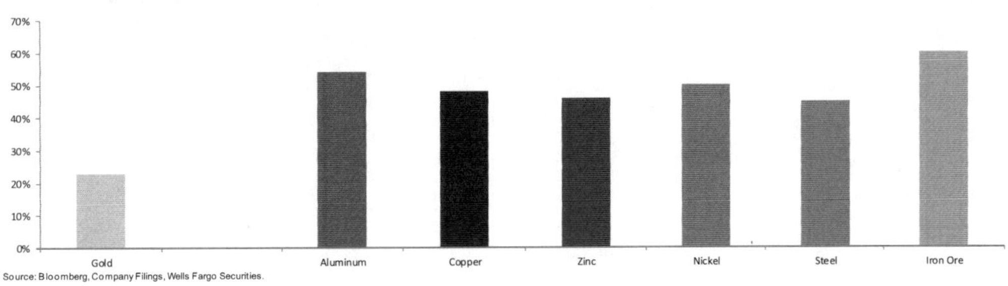

Source: Bloomberg, Company Filings, Wells Fargo Securities.

Source: Wells Fargo Securities and Bloomberg L.P.

Industrial China Showed Signs of Life in 2016

Chinese Fixed Asset Investment Growth (YoY)

Source: Bloomberg, Wells Fargo Securities.

Residential Construction Growth (YoY)

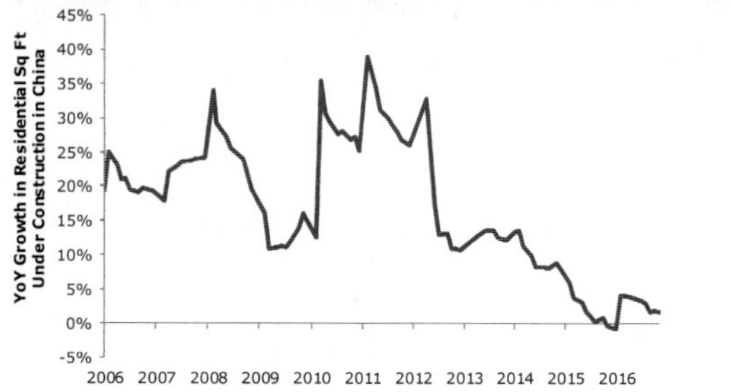

Source: Bloomberg, Wells Fargo Securities.

Chinese Electricity Consumption Growth (YoY)

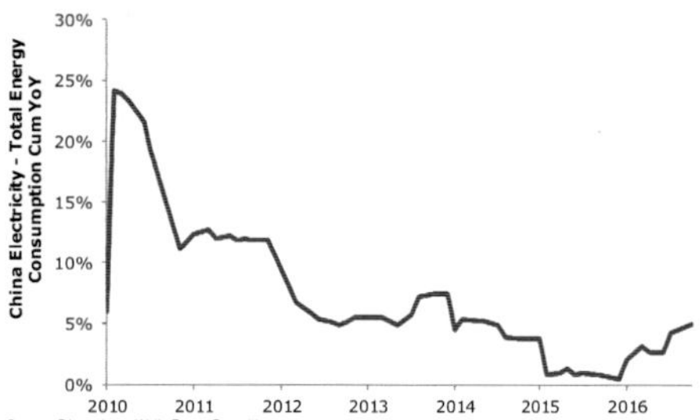

Source: Bloomberg, Wells Fargo Securities.

Chinese Passenger Car Sales

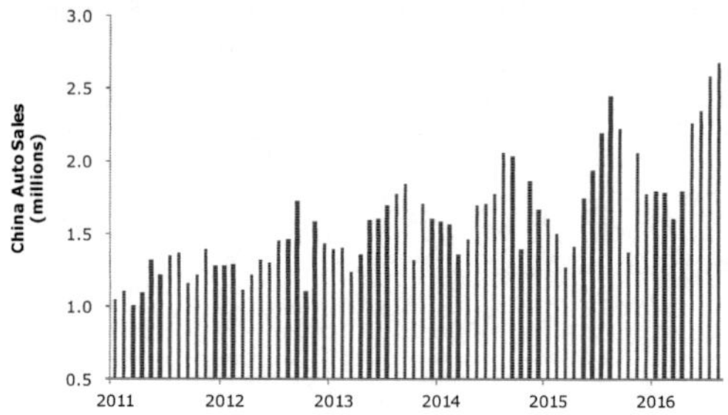

Source: Bloomberg, Wells Fargo Securities.

Wells Fargo Securities

Source: Wells Fargo Securities and Bloomberg L.P.

4

... As China Seems Willing to Inject Liquidity into System

RMB Loans into Chinese Economy

Source: Bloomberg, Wells Fargo Securities.

Wells Fargo Securities

Source: Wells Fargo Securities and Bloomberg L.P.

5

... But Signs of a Potential Pullback Are Starting to Appear

Commentary

- Following a year of robust spending fueled by large waves of fiscal stimulus in China, we are starting to see some signs of pulling back as we enter 2017.
  - The PBoC has begun to hike rates at the margin (MLF, SLF, Repos, etc.)
  - Local government debt issuance fell meaningfully in Q4 from the record levels experienced earlier in the year.
  - China's grid spending contracted sharply in late 2016.

Local Gov't Starts to Withdraw from Debt Markets

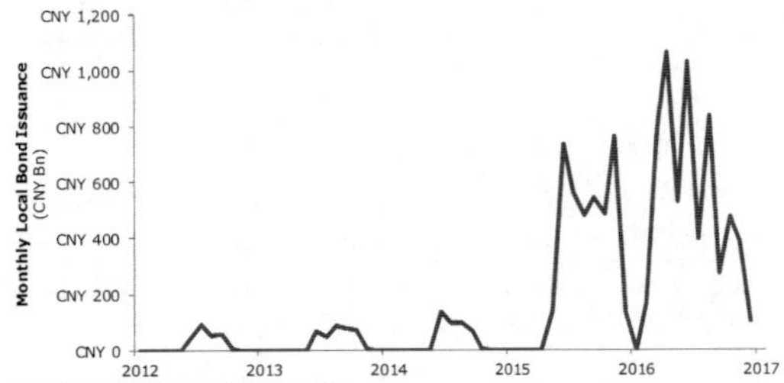

Source: Bloomberg, Wells Fargo Securities.

China's Grid Spending Fell in H2 2017

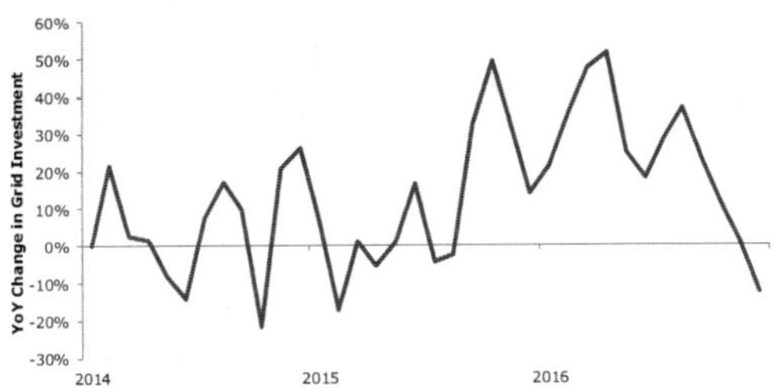

Source: Bloomberg, Wells Fargo Securities.

Falling CNY Supporting Industrial Cost Position

Source: Bloomberg, Wells Fargo Securities.

Source: Wells Fargo Securities and Bloomberg L.P.

Momentum Remains in Developed Economies

European Economy Continues to Trend Positively

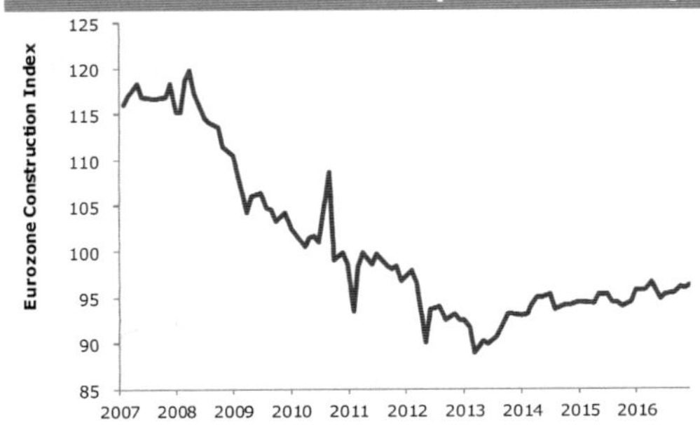

Source: Bloomberg, Wells Fargo Securities.

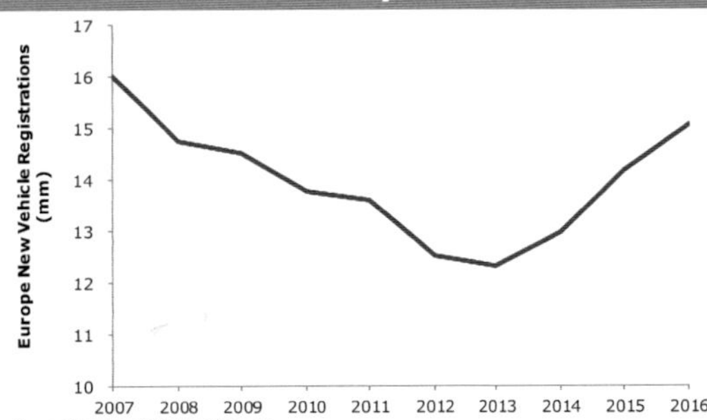

Source: Bloomberg, Wells Fargo Securities.

Trump Plans Could Spur Stagnant U.S. Zinc Consumption

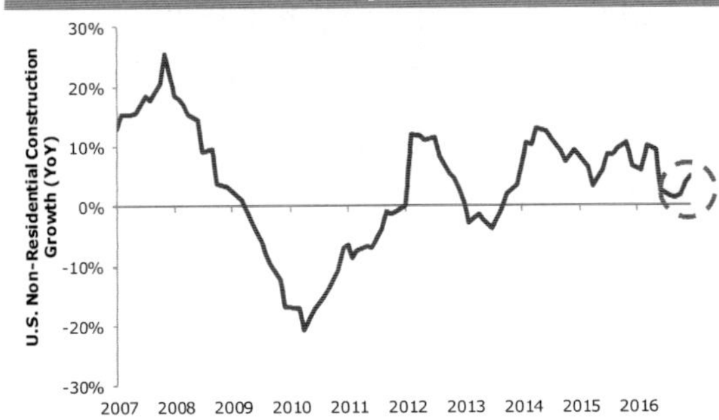

Source: Bloomberg, Wells Fargo Securities.

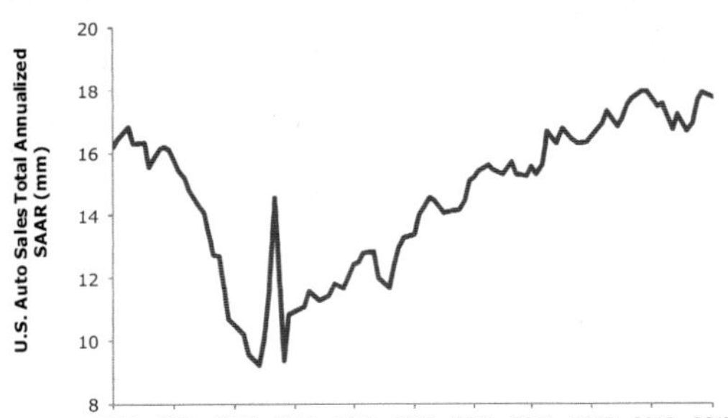

Source: Bloomberg, Wells Fargo Securities.

Source: Wells Fargo Securities and Bloomberg L.P.

Zinc Market Dynamics

Commentary

- Much like other base metals, zinc has experienced meaningful volatility of late as falling demand out of China has been met with lower prices.
- However supply has been impacted as well.
  - Major miners have pulled back production in the face of a weak price environment (Glencore).
  - Large mines have come to end of lives (Century, Lisheen).
- Despite rally in prices, underwhelming mine supply and low warehouse inventories support higher prices.

Major Supply Drivers

2015 → 2016 Supply Driver	Impact (mtpa)	2016 → 2017 Supply Driver	Impact (mtpa)
Century Mine (MinMetals)	-510	Dugald River Mine (MMG)	+160
George Fisher & Lady Loretta Mines (Glencore)	-245	Middle Tennessee Mines (Nyrstar)	+50
Lisheen Mine (Vedanta)	-175	Glencore Restarts	???
Golden Grove Mine (MinMetals)	-140		
McArthur River Mine (Glencore)	-135		

Source: Bloomberg, Company Reports, Wells Fargo Securities.

Supply Cuts Send Prices Higher

Source: Bloomberg, Wells Fargo Securities.

... But Inventories Remain Low

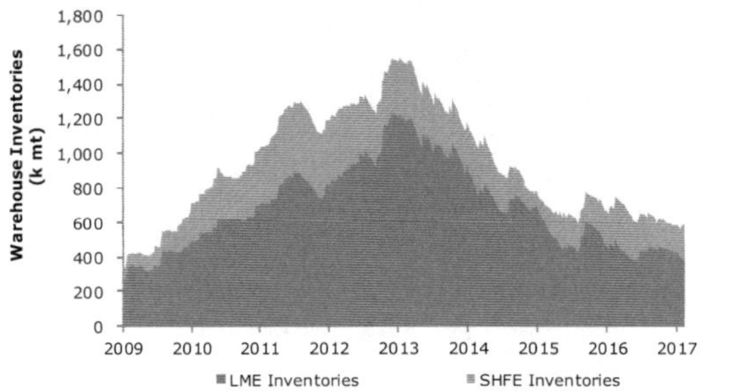

Source: Bloomberg, Wells Fargo Securities.

Source: Wells Fargo Securities and Bloomberg L.P.

What are the Credit Markets Telling Investors

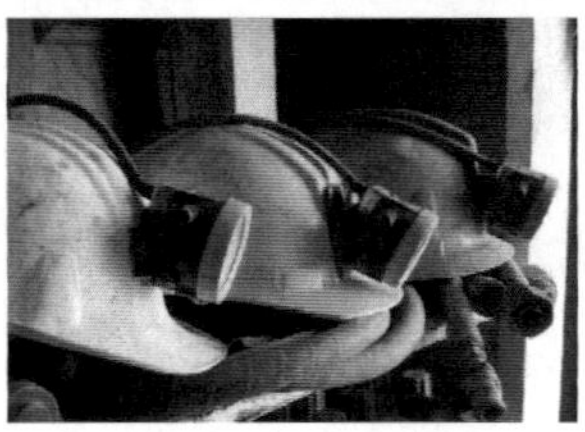

Metals & Mining Companies Delivered Growth with Help of Debt

Metals & Mining Producers Generated Record Profits on the Heels of Chinese Growth

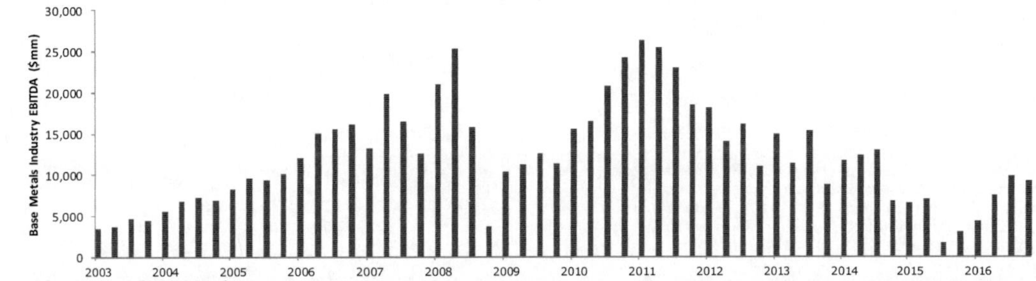

Source: Bloomberg, Company Filings, Wells Fargo Securities.

Debt Issuance Grew to Record Levels to Support Expansion

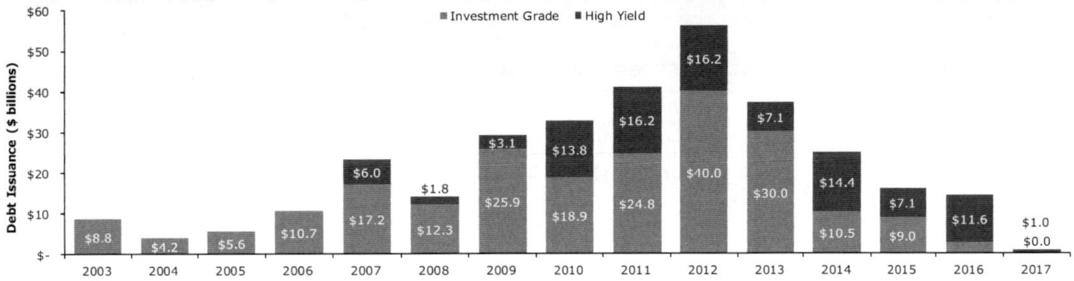

Source: Bloomberg, L.P., company reports and Wells Fargo Securities

China Fears, Commodity Fallout Jolted Capital Markets

Slowing Economy, Currency Devaluation, Stock/Real Estate Bubbles Among China Fears

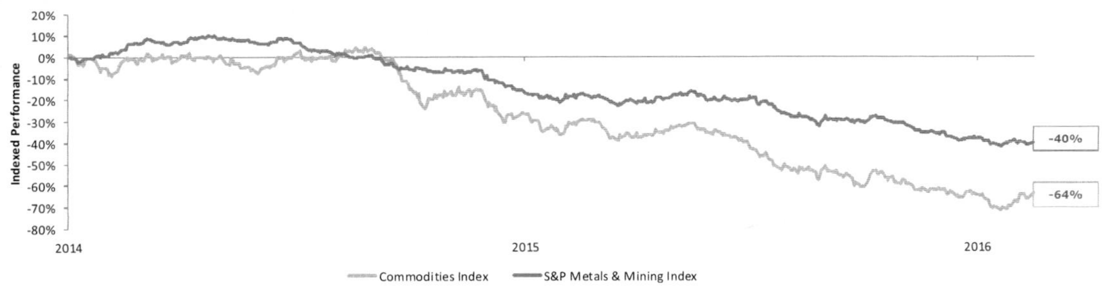

Source: Bloomberg, Wells Fargo Securities.

Debt Markets Respond, Limiting Access to Capital for Many Miners

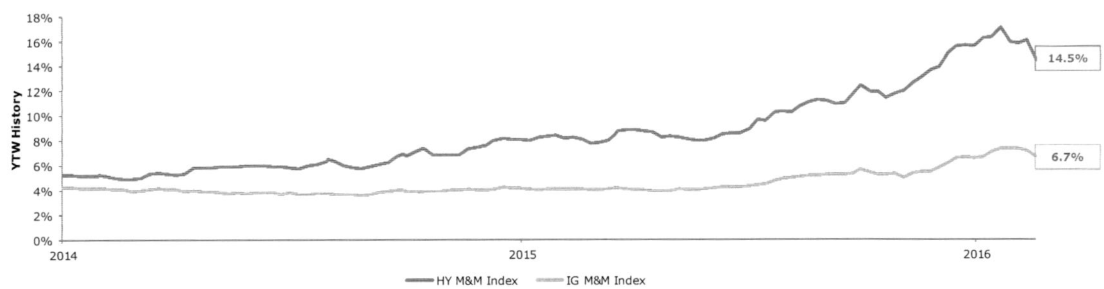

Source: Bloomberg, Wells Fargo Securities.

Capex Cuts & Improved Cost Profiles Helping Offset Low Prices

Commentary

- Metals and mining companies embarked on a path of self-preservation as it relates to production cost cuts and the easing of capital spending plans as a means of limiting cash burn amid persistent commodity price weakness, although this trends appears to have slowed due to the reopening of capital markets access and higher commodity prices.
- While it has not been a byproduct of the companies themselves, most foreign producers of metals are watching their cost profiles fall thanks to a wave of global currency devaluation against the U.S. dollar, helping to preserve margins as most metals are priced in USD.

Currency Devaluation Supporting Cost Positions

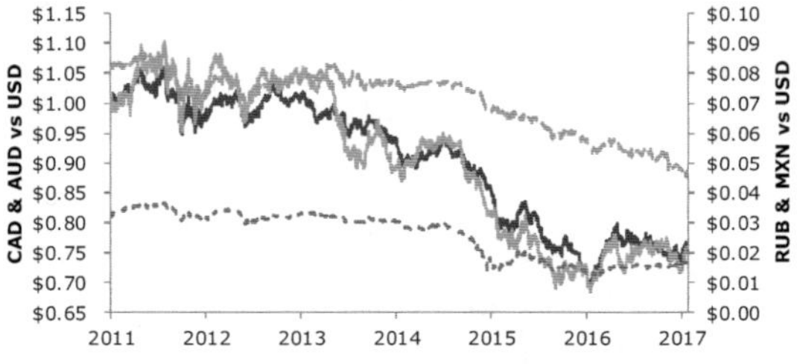

Source: Bloomberg, Wells Fargo Securities.

Companies are Slashing Capex to Weather the Storm

All figures in US$ millions	2010	2011	2012	2013	2014	2015	2016E
Diversified Miners	42,775	64,658	88,621	82,266	64,289	44,460	29,121
Precious Miners	10,947	18,915	24,135	20,454	13,556	10,119	8,985
Steel Producers	4,856	6,663	6,789	5,466	5,052	3,917	3,840
Coal Miners	2,575	3,887	4,080	2,997	2,494	573	588
Total	**61,152**	**94,124**	**123,625**	**111,183**	**85,390**	**59,069**	**42,534**
YoY % Change		**54%**	**31%**	**-10%**	**-23%**	**-31%**	**-28%**

Source: Company reports and Wells Fargo Securities' estimates

Self-Help Measures Looked to Shore Up Balance Sheet As Well

Divestitures & Equity Issuances Key to Bolstering Liquidity, Easing Market Fears

Actions Taken Since 2014

	AALLN	TECK	FCX	MTNA	FMCN
Asset Sales	✓ $1.9B	× n/a	✓ $11.0B	× n/a	✓ $0.9B
Equity Issuance	× n/a	× n/a	✓ $3.5B	✓ $3.1B	✓ $1.1B
Dividend Cuts	✓ Suspended	✓ Cut 90%	✓ Suspended	✓ Suspended	✓ Cut 95%
Debt Reductions	✓ $5.3B	✓ C$2.4B	✓ $4.4B	✓ $7.4B	× n/a

Source: Bloomberg, Company Filings, Wells Fargo Securities.

Commodity Prices Recover on Resurgent Chinese Demand

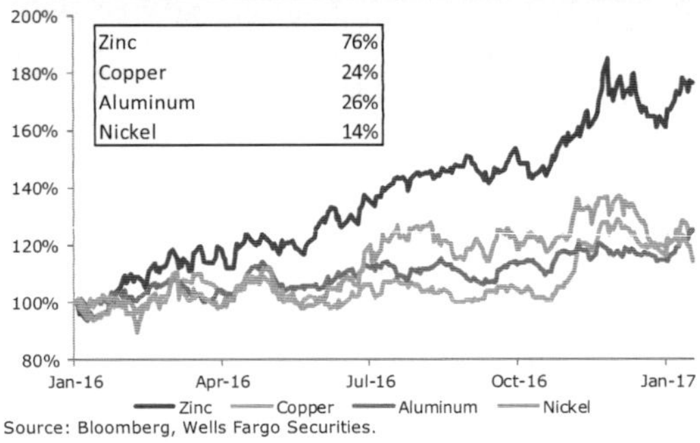

Source: Bloomberg, Wells Fargo Securities.

Company Actions, Improving Fundamentals Send Yields Back to Tights

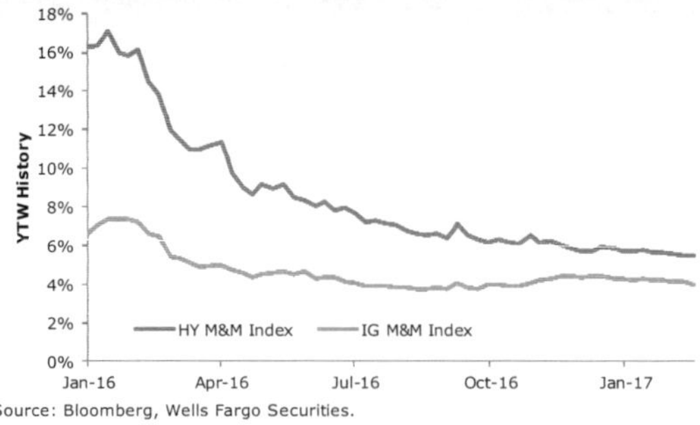

Source: Bloomberg, Wells Fargo Securities.

Case Studies: Teck Resources & Anglo American

Teck Resources: Price Recovery Allows for Substantial Debt Reductions

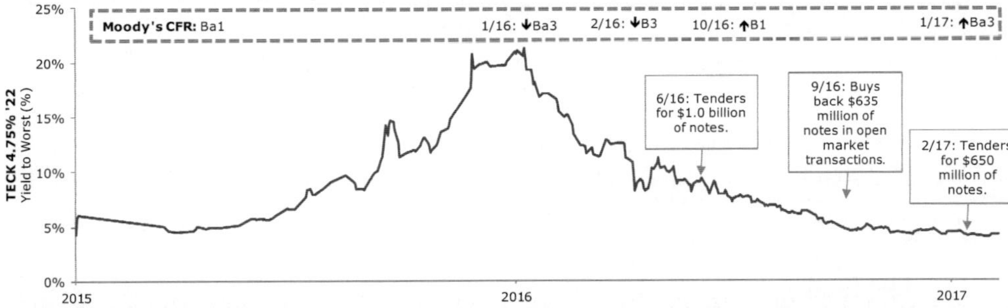

Source: Bloomberg, Wells Fargo Securities.

Anglo American: Asset Sales, Improved Cash Flows Sends Notes Back to Tights

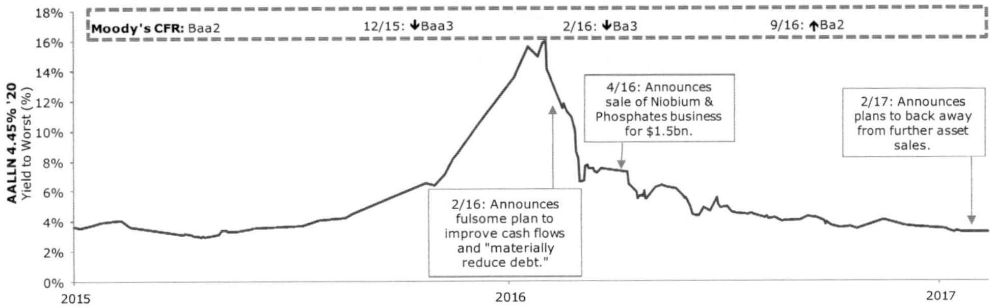

Source: Bloomberg, Wells Fargo Securities.

Wells Fargo Securities

Source: Wells Fargo Securities. and Bloomberg L.P.

14

Conclusions

Conclusions

- Following stimulus measures out of China, credit markets have reverted back to historical tights as investors have rebalanced portfolios towards more cyclical sectors.
- Barring any major change in Chinese policy, we would anticipate commodity markets to remain relatively balanced, providing some level of pricing support to commodity producers globally.
- We expect metals and mining companies to experience positive credit ratings momentum as ratings agencies re-rate producers following self-help measures and commodity price recovery.
- With access to capital renewed and ratings back on a positive trajectory, we expect companies to return to M&A markets and project development work in an effort to address looming metal deficits (zinc, copper, etc.).

Managed Slowdown in China Expected to Provide Sufficient Support to Debt Markets

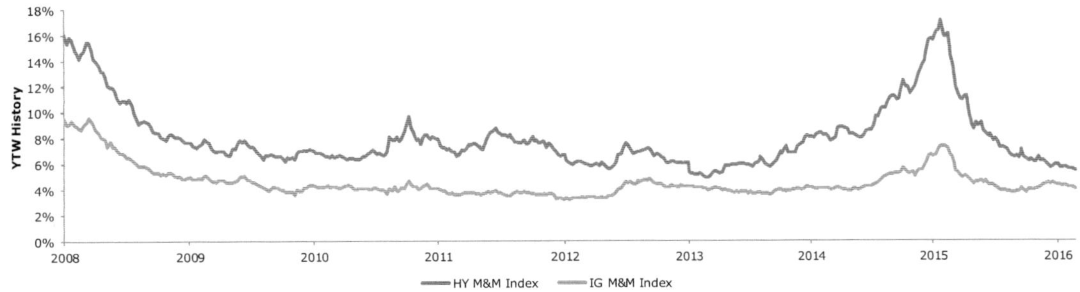

Source: Bloomberg, Wells Fargo Securities.

Wells Fargo Securities

Source: Wells Fargo Securities. and Bloomberg L.P.

15

Disclosures

Additional information is available on request.

This is a compendium report. To view current important disclosures and other certain content related to the securities recommended in this publication, please go to https://www.wellsfargoresearch.com/Disclosures or send an email to: wellsfargoresearch@wellsfargo.com or a written request to Wells Fargo Securities Research Publications, 301 South College Street 7th floor, Charlotte, NC 28202.

Analyst's Certification

The research analyst(s) principally responsible for the report certifies to the following: all views expressed in this research report accurately reflect the analysts' personal views about any and all of the subject securities or issuers discussed; and no part of the research analysts' compensation was, is, or will be, directly or indirectly, related to the specific recommendations or views expressed by the research analyst(s) in this research report.

Important Disclosures Relating to Conflicts of Interest and Potential Conflicts of Interest

Wells Fargo Securities does not compensate its research analysts based on specific investment banking transactions. Wells Fargo Securities research analysts receive compensation that is based on and affected by the overall profitability of their respective department and the firm, which includes, but is not limited to, investment banking revenue. Wells Fargo Securities may sell or buy the subject securities to/from customers on a principal basis or act as a liquidity provider in such securities.

Wells Fargo Securities Fixed Income Research analysts interact with the firm's trading and sales personnel in the ordinary course of business. The firm trades or may trade as a principal in the securities or related derivatives mentioned herein. The firm's interests may conflict with the interests of investors in those instruments.

Company Recommendation Definitions

Outperform: The bond's or issuer's spreads and/or returns are expected to exceed the spread performance or the return of the analyst's industry (or industry team's) coverage universe, on a risk-adjusted basis over the next 6-12 months. BUY

Market Perform: The bond's or issuer's spreads and/or returns are expected to be in line with the spread performance or the return of the analyst's industry (or industry team's) coverage universe, on a risk-adjusted basis over the next 6-12 months. HOLD

Underperform: The bond's or issuer's spreads and/or returns are expected to be below the spread performance or the return of the analyst's industry (or industry team's) coverage universe, on a risk-adjusted basis over the next 6-12 months. SELL

Sector Recommendation Definitions

Overweight The performance of the industry over the next 6-12 months is expected to outperform the relevant broad market benchmark.

Market Weight The performance of the industry over the next 6-12 months is expected to be in line with the relevant broad market benchmark.

Underweight The performance of the industry over the next 6-12 months is expected to underperform the relevant broad market benchmark.

About Wells Fargo Securities

Wells Fargo Securities is the global brand name for the capital markets and investment banking services of Wells Fargo & Company and its subsidiaries, including but not limited to Wells Fargo Securities, LLC, a U.S. broker-dealer registered with the U.S. Securities and Exchange Commission and a member of NYSE, FINRA, NFA and SIPC, Wells Fargo Prime Services, LLC, a member of FINRA, NFA and SIPC, Wells Fargo Bank, N.A. and Wells Fargo Securities International Limited, a U.K. entity investment firm authorized and regulated by the Financial Conduct Authority. The Wells Fargo Securities legal entity that takes responsibility for the production of the Product is the legal entity which the first named author is employed by. Non-US analysts may not be associated persons of Wells Fargo Securities, LLC. and therefore may not be subject to FINRA Rule 2242 restrictions on communications with subject company, public appearances and trading securities by the analysts, but will be subject to their own local regulatory requirements.

Disclosures

Notice to U.S. Investors

Unless prohibited by the provisions of Regulation S of the 1933 Act, this material is distributed in the U.S., by Wells Fargo Securities, LLC, which takes responsibility for its contents in accordance with the provisions of Rule 15a-6 and the guidance thereunder, under the U.S. Securities Exchange Act of 1934. Any transactions in securities identified herein may be effected only with or through Wells Fargo Securities, LLC.

Important Information for Non-U.S. Clients

EEA

The securities and related financial instruments described herein may not be eligible for sale in all jurisdictions or to certain categories of investors. For recipients in the EEA, Wells Fargo Securities International Limited ("WFSIL") disseminates Research which has been approved for the purposes of Section 21 of the Financial Services and Markets Act 2000 ("the Act"). WFSIL is a U.K. incorporated investment firm authorized and regulated by the Financial Conduct Authority. For the purposes of Section 21 of the Act, WFSIL does not deal with retail clients. The FCA rules made under the Financial Services and Markets Act 2000 for the protection of retail clients will therefore not apply, nor will the Financial Services Compensation Scheme be available. This report is not intended for, and should not be relied upon by, retail clients.

Australia

Each of Wells Fargo Securities, LLC, Wells Fargo Securities Asia Limited and Wells Fargo Securities International Limited is exempt from the requirements to hold an Australian financial services license in respect of the financial services it provides to wholesale clients in Australia. Wells Fargo Securities, LLC is regulated under U.S. laws, Wells Fargo Securities Asia Limited is regulated under Hong Kong law, and Wells Fargo Securities International Limited is regulated under U.K. law, all of which differ from Australian laws. Any offer or documentation provided to Australian recipients by Wells Fargo Securities in the course of providing the financial services will be prepared in accordance with the laws of the United States, Hong Kong or U.K. and not Australian laws.

Hong Kong

This report is issued and distributed in Hong Kong by Wells Fargo Securities Asia Limited ("WFSAL"), a Hong Kong incorporated investment firm licensed and regulated by the Securities and Futures Commission to carry on types 1, 4, 6 and 9 regulated activities (as defined in the Securities and Futures Ordinance (Cap. 571 of The Laws of Hong Kong), "the SFO"). This report is not intended for, and should not be relied on by, any person other than professional investors (as defined in the SFO). Any securities and related financial instruments described herein are not intended for sale, nor will be sold, to any person other than professional investors (as defined in the SFO). The author or authors of this report is or are not licensed by the SFC. Professional investors who receive this report should direct any queries regarding its contents to Mark Jones at WFSAL (email: wfsalresearch@wellsfargo.com).

Japan

This report is distributed in Japan by Wells Fargo Securities (Japan) Co., Ltd, registered with the Kanto Local Finance Bureau to conduct broking and dealing of type 1 and type 2 financial instruments and agency or intermediary service for entry into investment advisory or discretionary investment contracts. This report is intended for distribution only to professional investors (Tokutei Toushika) and is not intended for, and should not be relied upon by, ordinary customers (Ippan Toushika).

The ratings stated on the document are not provided by rating agencies registered with the Financial Services Agency of Japan (JFSA) but by group companies of JFSA-registered rating agencies. These group companies may include Moody's Investors Services Inc, Standard & Poor's Rating Services and/or Fitch Ratings. Any decisions to invest in securities or transactions should be made after reviewing policies and methodologies used for assigning credit ratings and assumptions, significance and limitations of the credit ratings stated on the respective rating agencies' websites.

Disclosures

Distribution of Recommendations

As of 2/27/2017

31.6% of issuers and securities covered by Wells Fargo Securities Fixed Income Research are rated Outperform.	Wells Fargo Securities has provided investment banking services for 23.2% of its Fixed Income Outperform-rated issuers.
50.7% of issuers and securities covered by Wells Fargo Securities Fixed Income Research are rated Market Perform.	Wells Fargo Securities has provided investment banking services for 26% of its Fixed Income Market Perform-rated issuers.
17.7% of issuers and securities covered by Wells Fargo Securities Fixed Income Research are rated Underperform.	Wells Fargo Securities has provided investment banking services for 17.3% of its Fixed Income Underperform-rated issuers.

Published reports and additional disclosure information are available at www.wellsfargoresearch.com.

Source: Wells Fargo Securities. and Bloomberg L.P.

驰 宏 锌 锗
CHIHONG Zn&Ge

YUNNAN CHIHONG Zn & Ge CO.,LTD

QU JING YUNNAN CHINA

GALLOP THE WORLD LOFTY AMBITIONS

Yongguang Luo

General manager

OUTLINE

1	Zinc production capacity of China in 2016
2	Development of Chihong company
3	Industry layout of mining and metallurgy
4	Smelting enterprise layout
5	Smelting equipment manufacturing

驰 宏 锌 锗
CHIHONG Zn&Ge

1

Zinc production capacity of China in 2016

驰 宏 锌 锗
CHIHONG Zn&Ge

Production of scale enterprise

Region	Production
HUNAN	715225
HENAN	270701
LIAONING	274400
YUNNAN	843795
NEIMENGGU	571736
SICHUAN	213859
SHAANXI	781167
GUANGXI	390430
GUANGDONG	201964
GANSU	291598
QINGHAI	38455
JIANGXI	98007
ANHUI	100139
Summation	4791476

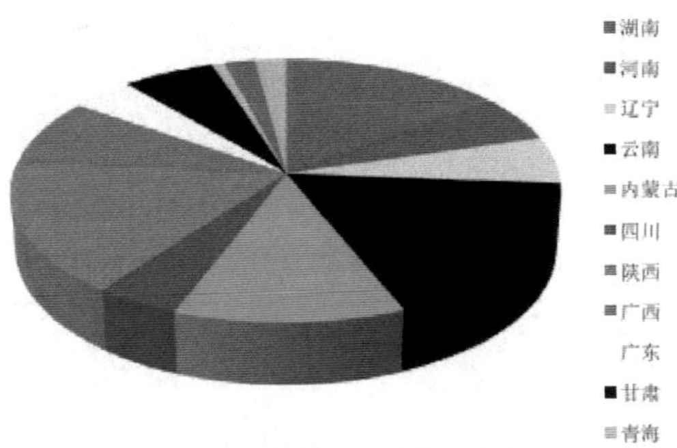

Regional production

Production of scale enterprise

Enterprise	Output	Inc. or Dec. (%)	Enterprise	Output	Inc. or Dec. (%)
Zhuzhou Smelter	486,552	-6.6	Zijin Mining	214,000	-2.2
Hanzhong Zinc Industry	362,565	7.9	Shaanxi Zinc Industry	212,900	4.2
Southern Nonferrous Metallurgy	324,900	6.2	Dongling Zinc Industry	205,702	-10.6
Yunnan Chihong Zn & Ge	287,354	15.8	Zhongjin Lingnan	201,964	-1.0
Huludao	274,400	2.9	Yunnan Copper and Zinc Industry	162,000	12.5
Yuguang Gold & Lead	264,201	-13.66	Inner Mongolia sing-an copper zinc smelter	113,000	-3.0
Baiyin Nonferrous Metals Company	252,098	-1.2	Tongling Metal Group	100,139	22.1
Chifeng Smelter	217,936	1.3	Summation	3679,711	

Scale enterprise production accounted for 76.80% of total output

The growing lead and zinc smelter in China

In order to obtain stable raw material guarantee, many lead and zinc smelter now implement the resource integration strategy, starting mass in acquisition, merger, domestic holding mining enterprises and overseas expansion strategy of mineral resources.

Mining projects in Canada, Bolivia and Peru, owned by YUNNAN CHIHONG Zn & Ge CO.,LTD.

Zhongjin lingnan acquired PEM mining company.

Minmetals acquired the OZ mining company with abundant capital strength.

China's lead and zinc smelting project gradually develop from a single smelting production to whole industry chain including "mining- beneficiation- smelting- deep processing".

The current situation of lead and zinc smelting in China

China's lead and zinc in the majority with private smes, scattered industry concentration degree is low, unreasonable industry structure, enterprise competitive advantage is not obvious. More than 320 lead and zinc smelting companies distribute in 30 provinces.

Small and medium-sized enterprise equipment level is low, smelting technology lag behind, low level of smelting, fire still account for 30%, serious environmental pollution, resource comprehensive utilization rate is not high, cause contradiction of industrial sustainable development and ecological environment protection.

The current situation of lead and zinc smelting in China

Rising costs, increased capacity, and product price fluctuations, the consumer market atrophy, lead and zinc smelting industry in our country entered the benefits of being marginalized, the industry will face losses.

Resources, external dependence enhances unceasingly.Lead and zinc resources in China is small mines, lean ore, difficult mining difficult dressing quantity is big, the scarcity of available resources.

The problem of lead and zinc smelting project construction in China

Problem (1)	**Relative to foreign counterparts in terms of the project life cycle is short, namely the service cycle is short.**
Problem (2)	**Severe environmental conditions restrict the survival and development of the project, technology innovation blind competition resulting in a decline in the profitability and overall competitive strength of the project.**
Problem (3)	**China's lead and zinc smelting projects lack of mineral resources, mineral resources optimized conformity and smelting projects don't match.**

The situation of typical Chinese zinc smelter

➢Chifeng Smelter

Chifeng Zinc Industry Co., Ltd. controlled by China Nonferrous Metal Construction Co., Ltd., South Africa Cooper companies and other state-owned enterprises for sino-foreign joint ventures, the total assets of 3.5 billion. The main products are zinc ingot and zinc alloy of 46 brands. With an annual output of 210,000 tons of zinc products, of which the total output of zinc alloy 48%.

Year	2012	2013	2014	2015	2016
Production of Zinc Products (Ten thousand tons)	20.6	21	21.7	21.5	21.7
Total Profit (Ten thousand yuan)	20	1920	1103	30	13000

The situation of typical Chinese zinc smelter

Yuguang Gold & Lead

Year	2012	2013	2014	2015	2016
Production of Zinc Products (Ten thousand tons)	22.5	24	24.5	26	26.5
Total Profit (Ten thousand yuan)	2900	4200	13000	12000	9200

Henan yuguang XinYe co., LTD to produce zinc ingots, indium ingot, sulfuric acid and other non-ferrous metal and chemical products. In March 2007, the company production of "YG" brand zinc ingots registered in Shanghai futures exchange. In June 2008, yuguang brand high purity zinc registered in London. The current production capacity of 260000 tons of electrolytic zinc, alloy 100000 tons, 410000 tons of sulfuric acid.At the same time, comprehensive recovery of lead, silver, copper, cadmium, cobalt, indium, iron and so on more than 10 kinds of valuable metals.

2

Development of Chihong company

驰 宏 锌 锗

CHIHONG Zn&Ge

Development of Chihong company

Yunnan Huize Lead-Zinc mine, founded in January 1951, is the earliest enterprise to extract Ge from lead-zinc oxide ore for advanced national defense industry.

It has made contributions to the successful development of “two bombs and one satellite”.

Enterprise history

July 18, 2000, five units co-sponsored the establishment of Yunnan Chihong Zn & Ge Co., Ltd..

The registered capital for Chihong Zn & Ge is about 1,667,560,890 yuan; by Yunnan Metallurgical Group holding 49.74% and the headquarters is located in yunnan qujing city economic and technological development zone.

Enterprise history

April 20, 2004, Yunnan Chihong Zn & Ge Co., Ltd. is listed on Shanghai Stock Exchange and become a-share listed companies.

Development of the company

Productivity scale	With the actual production capacity of 3 million tons of mining, mineral processing design production capacity of 4.5 million tons, lead and zinc smelting 650, 000 tons, lead and zinc production scale ranks first in Yunnan Province, the national third. The main equipment, environmental protection, occupational health and technology reached the domestic first-class, the international advanced level, lead-zinc resources reserves, profitability in the same industry forefront.
Industry scale	The company covers a wide range of industries including geological exploration, mining, mineral processing, metallurgy, chemical industry, deep processing trade and scientific research and has 39 subsidiaries in Yunnan, Sichuan, Inner Mongolia, Heilongjiang, Tibet, Hong Kong, Canada, Australia, Bolivia and so on. Varieties of resources cover the Pb, Zn, Cu, Mo, Au, Ag, Ge and other non-ferrous metals
Scale of operation	In 2014, the total assets of the company is 33.5 billion yuan and the net assets is 11.5 billion yuan and the "Fortune China" 500 ranked 269, 271 and 292 for 2014, 2015 and 2016, respectively.

Development of the company

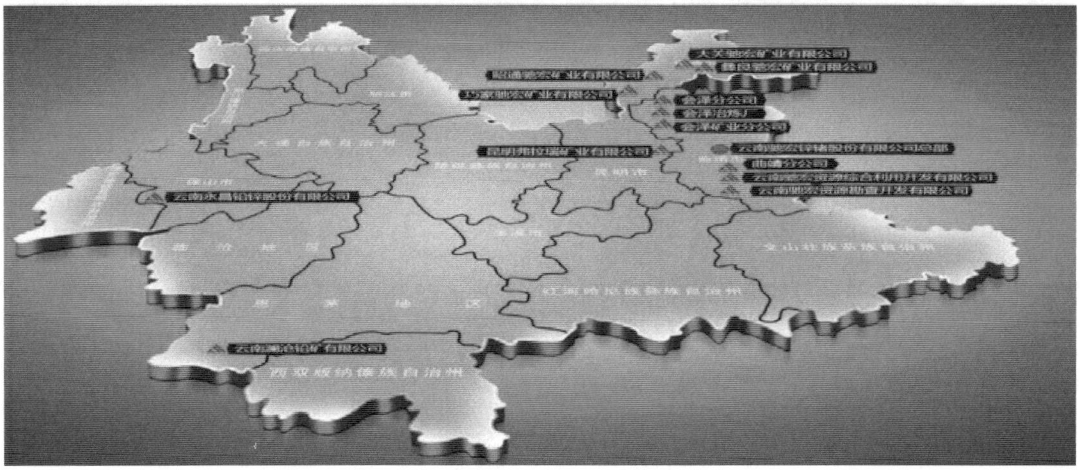

Yunnan's industrial layout

Development of the company

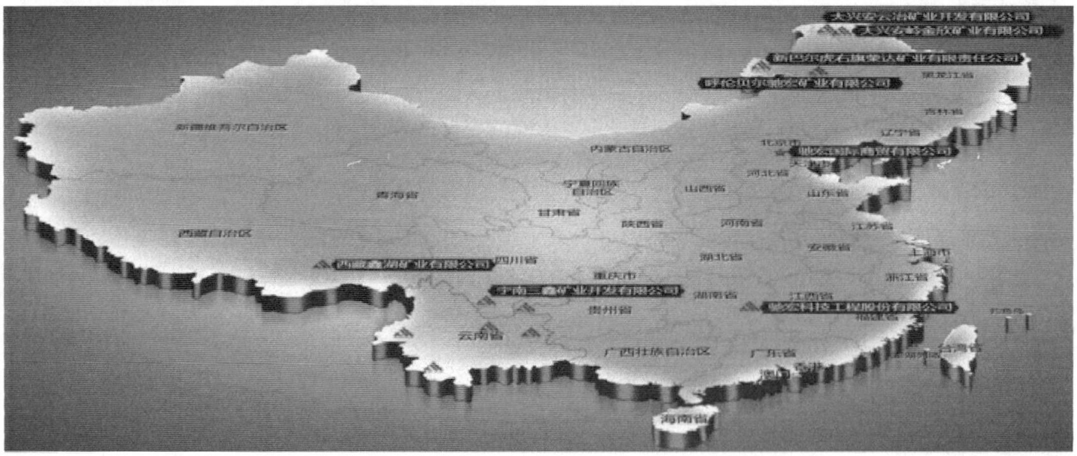

China's industrial layout

Development of the company

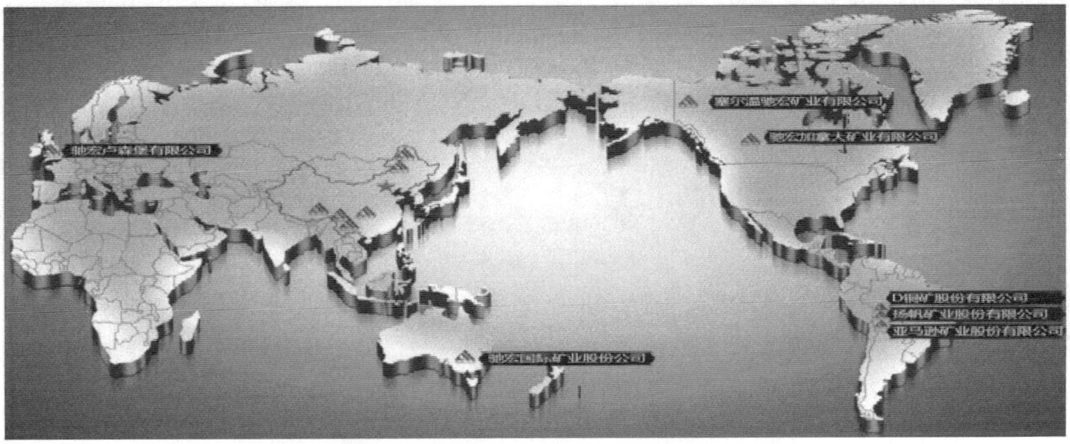

World industrial layout

Main products of the company

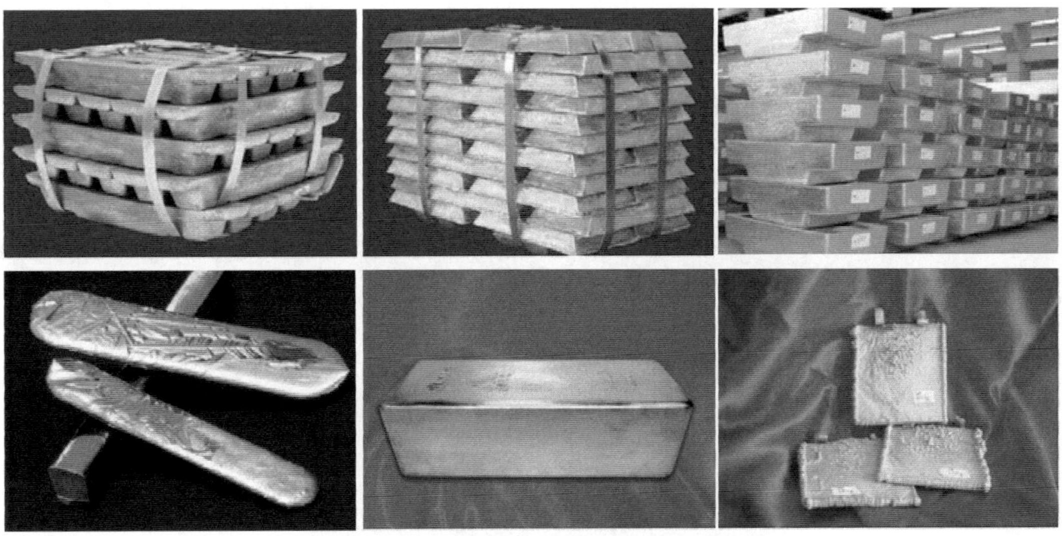

Main products of the company

Registered trademark

銀晶牌

银鑫牌

金沙 JIN SHA

华达

驰宏锌锗 CHIHONG Zn&Ge

3

Industry layout of mining and metallurgy

驰宏锌锗 CHIHONG Zn&Ge

Mine article

Chihong company has 15 mines (sub) companies, involving Yunnan, Guizhou, Sichuan, Inner Mongolia, Heilongjiang, Tibet, mining right 122 and covers an area of nearly 2,700 square kilometers.

Mine article

Domestic prospecting results

Chihong company invests 5.781 billion yuan for resource integration. Through the exploration and integration of resources accumulated reserves of Pb 575 million tons, Zn 7.45 million tons, Cu 310,000 tons, Ag 5401 tons, Ge 490 tons, Au 4.8 tons, Mo 246 tons. At present, domestic reserves of 333 and above Ga reserves of 40,000 tons, Re 55 tons, lignite 2.2 million tons; owned by the world's largest single Mo resources and the potential value is about 560 billion yuan. In the Yunnan mining rights, there are 28.6 million tons of phosphate rock and gypsum more than 3.2 million tons.

Mineral	Pb	Zn	Ge	Ag	Cu	Mo	Ga	P	Au	Re	Coal
Storage	575	745	0.049	0.54	31	246	4	2680	0.0048	0.055	220

Foreign prospecting results

Chihong company maintained in the domestic advantage of non-ferrous metal resources, at the same time, it is actively in layout of the global resource strategy and to develop overseas resources market, foreign companies involved in Luxembourg, Canada, Australia, Bolivia and other countries. By the end of December, foreign reserves of 333 and above reserves of Pb-Zn resources 26.89 million tons, Cu 530, 000 tons, Au 12.31 tons, Sb 440, 000 tons, converted into zinc equivalent of 30 million tons. The potential value of resources is 450 billion yuan.

Huize Mining Company

The Huize mining company is the representative of the digital mine. It is the largest shaft to enhance the capacity in southwest and the mining deepest non-ferrous metal mines in China.

Mining Technology

Paste filling technology is a resource-saving, environmental protection, production safety in one of the leading domestic and international filling technology. In 2008, it is awarded the "national key environmental protection practical technology demonstration project" by China Environmental Protection Industry Association.

4

Smelting enterprise layout

驰 宏 锌 锗
CHIHONG Zn&Ge

Zinc smelting process and production capacity

Company	Technology	Production capacity	Quantity of processing ore	Notes
Qujing Resource Utilization Company	Two leaching + smoke furnace	120,000	ZnS mine:260,000	Mechanical automatic stripping of zinc
Huize Smelting Company	Two leaching + smoke furnace Zinc oxide recovery	150,000	ZnS mine:200,000 ZnO mine:110.000	
Hulunbeier Chihong Mining Company	Oxygen pressure leaching	150.000	ZnS mine:310,000	

Major economic and technical indicators

Major economic and technical indicators

Indicators	Unites		
Zinc recovery	%	99.2	
Consumption of zinc powder	Kg/t	50	grade：Zn=80-85%
DC power consumption	Kwh/t	3160	Counting from the mother liquid
Energy consumption	kgcet	880	
		1060	
Efficiency	%	≥91	48 hours

Technical research direction

Our main research direction:
Purification of organic matter and balancing of magnesium

ion in zinc sulfate solution:
Using proprietary oxidation-hydrometallurgy technology.

Scattered elements (Ga, In, Tl, Se, Te) in zinc sulfate solution : Using proprietary technology of multi-element zinc powder (more than 5 yuan) to ensure the contents of Ga, In, Ti, and Te dropping to less than 1mg / L

Effect of organic compounds on zinc electrodeposition

Table2 Experimental results of effect of organic compounds on zinc electrodeposition

Date	09.11.08	09.11.10	09.11.12	09.11.14	09/.116	09.11.18	09.11.20	09.11.22	09.11.24
COD (mg/L)	30.69	58.87	59.68	60.49	62.10	66.29	75.12	76.74	79.16
Effeciency (%)	91.93	84.79	79.28	75.81	72.22	70.92	65.23	60.12	55.36
DC consumpition	2858.8	3313.3	3495. 7	3564.60	3833.89	4030	4211.3	4610	5021
Pb slice (%)	0.0009	0.0017	0.0028	0.0041	0.0048	0.0056	0.0067	0.0078	0.012

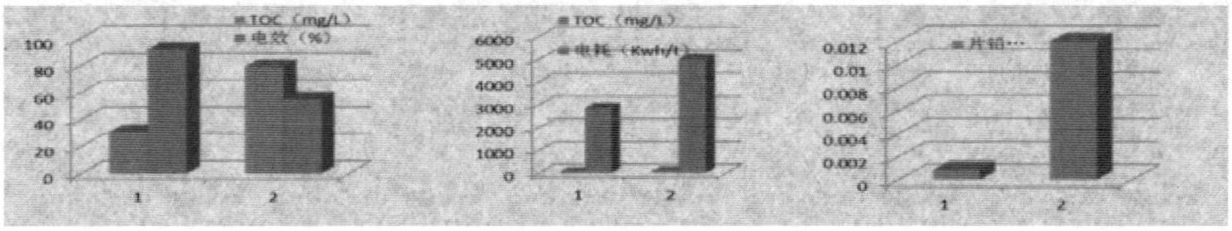

The results show that increasing organic matter (TOC characterization) from 30.69mg / L to 79.16 mg / L in solutions, the electrodeposition efficiency of zinc will reduce from 91.93% to 55.36% in 48 hours, and the DC consumption increases from 2858.8 Kw · h / t to 50212Kw · h / t per slice, with content of lead slice from 0.0009% to 0.012%.

Characteristics of zinc production process

1、Organic matter purification and magnesium ion balance in zinc sulfate solution

The company has a patented technology, which can effectively remove the organic matter in the solution of zinc sulfate solution to ensure that the TOC ≤40mg/L、Mg^{2+}≤12g/L。

2、Treatment technology of zinc oxide dust

The day processing is 380 tons of zinc oxide dust at Huize smelting company, use the seven purification and purification the organic matter, germanium and scattered metal elements in solution. The product is 100% 0# zinc.

Effect of Mg^{2+} on zinc electrodeposition

Table 2 Effect of level of Mg^{2+} on electrodeposition efficiency

Number	Mg^{2+} (g/L)	Efficiency (%)	Tank voltage (V)	DC power consumption (Kw·h/t)
1	5.2	91.83	3.24	2893
2	9.8	91.28	3.28	2947
3	15.2	89.80	3.40	3105
4	19.9	88.68	3.45	3190
5	25.3	86.52	3.49	3308
6	30.4	84.47	3.52	3446
7	34.9	82.53	3.58	3577

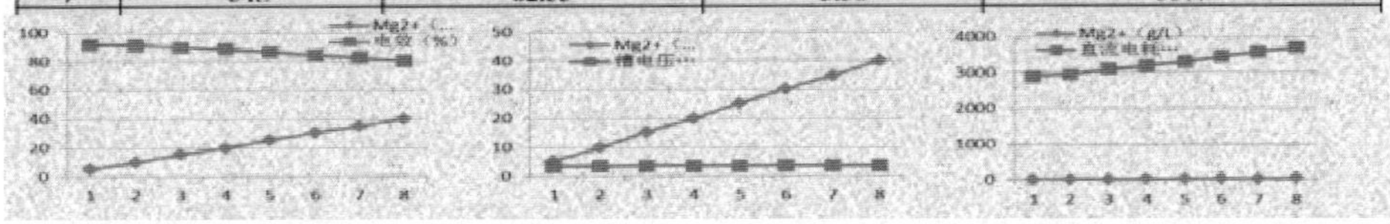

The results show that increasing level of Mg^{2+} to 40.2g / L from 5.2g / L , will result in the electrodeposition efficiency dropping, from 91.83% to 80.26% in 48 hours, tank voltage rising from 3.24V to 3.61V, and DC consumption increasing from 2893Kw · h /t to 3688Kw · h / t per slice.

Industrial application production site

Technology and Equipment Level

Smoke furnace system

Large handling capacity: smoke furnace can handle a variety of lead and zinc smelting slag, the annual handling capacity is 150,000 tons, it can realize continuous converting using all cold material

Long life of water jacket: through independent research and development on the flue gas outlet of smoke furnace, one or two layers of water jacket structure and material are optimized. The water jacket can run continuously for ten months.

Integration of fuming furnace and waster boiler: use the "smelting flue gas waste heat recovery-waste heat power generation technology" effective recovery of waste heat

Smelting waste residue harmless: after converting process of the high temperature furnace slag quenching by quenching water, it can be used in building materials industry and effectively reflects the circular economy features of "resource, reduction and reuse".

Smelting waste gas treatment

Environment priority investment

Investment

= Industry average140%

Technology + Equipment + Control = **National control standards ½ (≤150ppm)**

- **Technology:** High efficiency power, double turn and double suction technology, independent optimization of catalyst filling technology
- **Equipment:** Imported key equipment: catalyst, conversion fan, heat exchanger, system control key valves, instrumentation
- **Control:** Key operating parameters of control system, control parameters with automatic alarm device.

Smelting wastewater treatment

Investment 275 million yuan to build the reuse project of industrial wastewater, domestic sewage and initial rainwater. The industrial wastewater, domestic sewage and initial rainwater are recovery storage and will reuse after sewage treatment. The water of city sewage treatment plants is employed into the factory as a production water and to achieve the zero emissions of industrial wastewater.

5

Smelting equipment manufacturing

驰 宏 锌 锗
CHIHONG Zn&Ge

Equipment manufacturing

- At present, the company has its own research and development of automatic stripping machine, independent production of a variety of matching automatically stripped for cathode and anode plate
- The joint manufacturing anode crane automatically is proceeding.

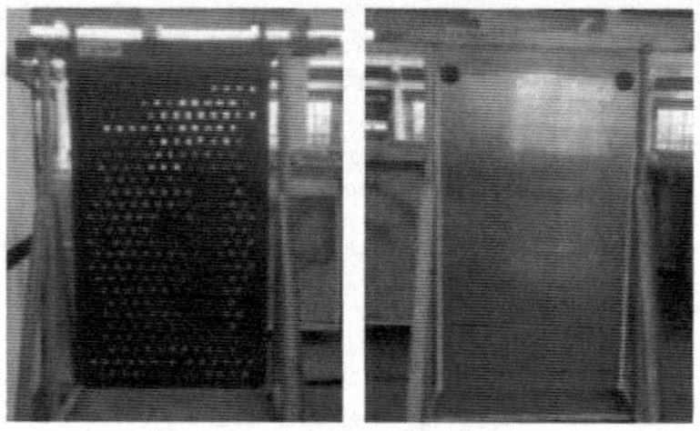

3.2m² electrodes

Equipment manufacturing

Cathode stir welding equipment

Cathode stir welding equipment

Cathode insulating molding equipment

Cathode stir welding equipment

Anode beam casting equipment

Anode rolling and welding equipment

Anode punching blanking equipment

Anode stir welding equipment

Independent production of a variety of matching automatically stripped for cathode and anode plate

Thank you for your attention and support

We sincerely look forward to cooperation with you

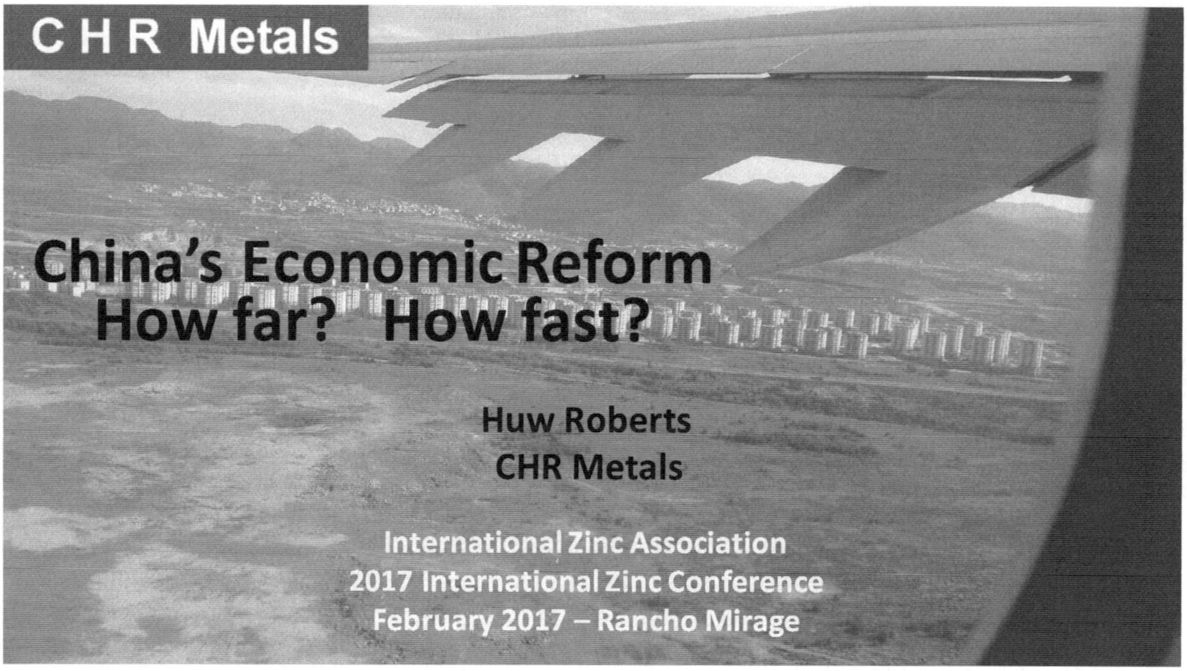

C H R Metals

Why is economic reform necessary?

- Huge economic success over past 25 years is not sustainable under current growth model
- Export, then investment-led growth has multiplied per capita GDP, but not without significant costs
- Cost to the environment is now obvious to all but rising income inequality and debt levels pose social and financial risks
- Export-driven growth before global financial crisis of 2008 generated large external imbalances
- Investment-led growth post 2008 has been heavily reliant on credit and often inefficient, with misallocation of capital and large internal imbalances generated
- Desired shift is from traditional manufacturing and heavy industry to tech-driven "new economy" and expanding consumer demand

February 2017 Slide 2

CHR Metals — China's growth on strong trajectory

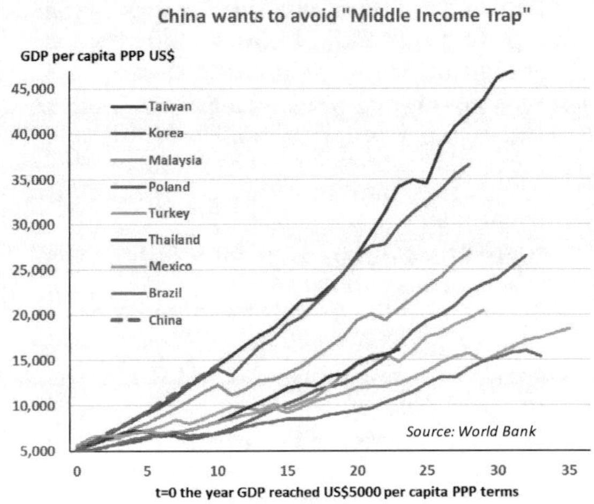

- For the past 10 years China's GDP per capita has been on the same path as Taiwan & Korea
- It has outpaced other typical emerging/ industrialising economies
- But massive investment spending and reliance on credit to fuel growth pose significant risks
- Will China fall into the so-called middle income trap?
- China's dash for growth mimics the experience of Taiwan and Korea...
- ...but these are much smaller countries with much smaller populations, economies and impact on global markets

February 2017 Slide 3

CHR Metals — Key drivers of growth are changing

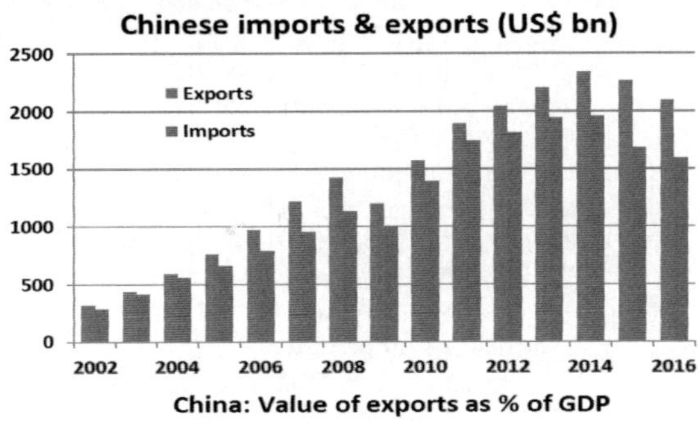

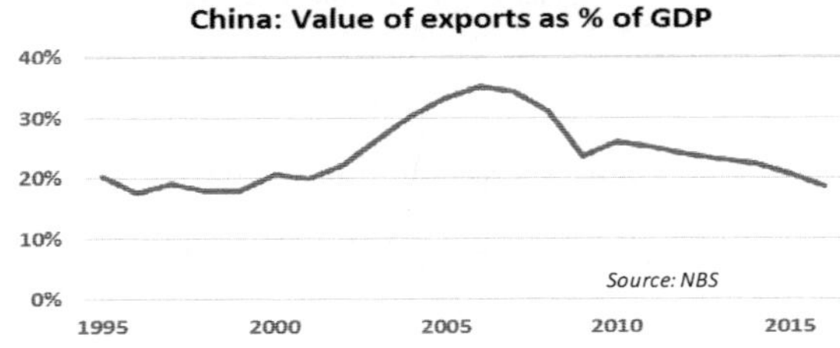

- Value of China's exports (in US$) peaked in 2014, as did imports
- Decline in 2015 and 2016 reflects weaker global trade overall, as well as competitive pressures on Chinese exporters
- Domestic costs are rising – wages, land, environmental costs, etc.
- Value of exports as per cent of GDP peaked at 35% in 2006, but is now back to level seen pre-WTO accession
- Exports remain important but no longer an engine of growth for China
- China is no longer competitive as the "world's factory"

February 2017 Slide 4

C H R Metals

China's position in world trade

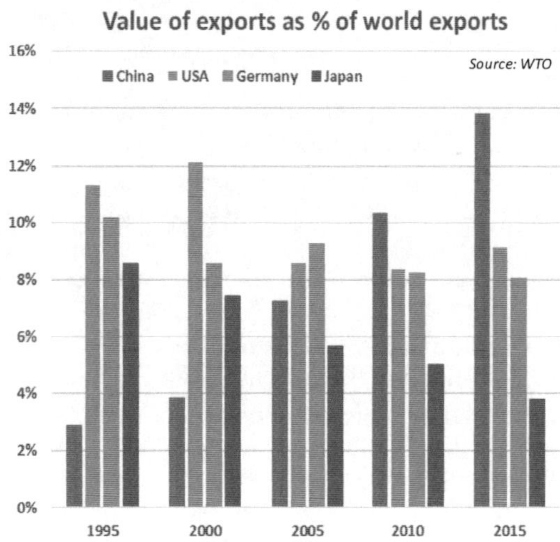

- Despite falling in absolute terms and as a per cent of GDP, Chinese exports are still a major factor in world trade
- China's relative gains have largely been at the expense of Japan, USA and various European economies – the mature economies - and also reflected in their use of metals...
- ...but China's expansion into world markets has squeezed emerging/industrialising economies
- These countries have seen minimal or no growth in their share of world trade over the past decade
- This is also reflected in little or no overall growth in use of some metals outside China in recent years

February 2017 Slide 5

C H R Metals

How China's economy is changing

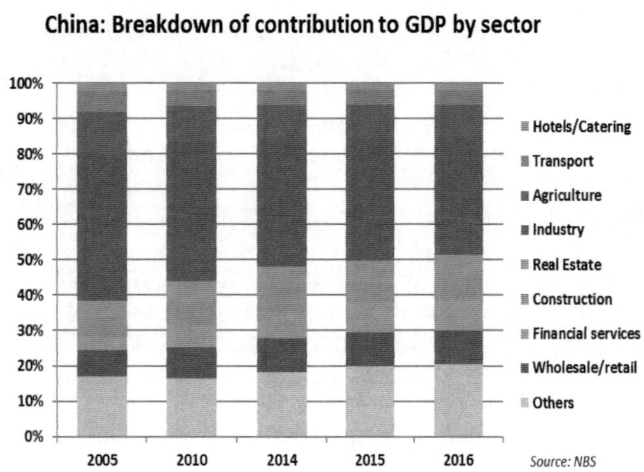

- The contribution of industry to China's GDP has clearly shrunk over the past decade but still accounts for a third of economic output
- Contribution of agriculture is also lower
- Major gain has been in financial services, from 4% in 2005, to over 8% last year
- Growth also in the contribution of wholesale & retail trades and "other"
- Other category includes
  - Public Management, Social Security & Social Organisation
  - Education
  - Information Transmission, Software & IT
  - Leasing & business services
  - Health & Social service
  - Others
- Latest detailed breakdown of Others category is for 2014

February 2017 Slide 6

C H R Metals — Investment patterns show little change

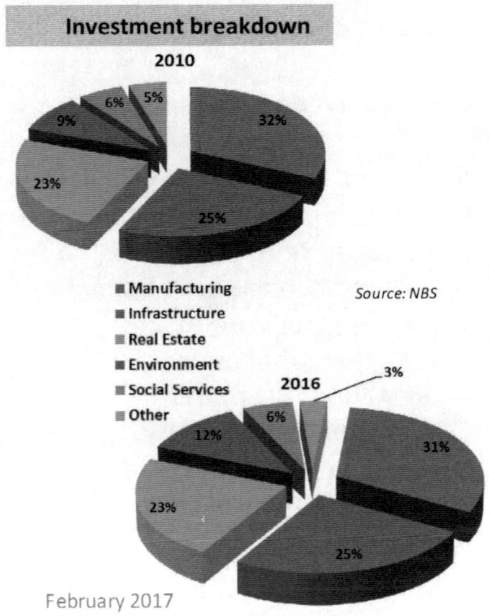

- Investment in manufacturing capacity accounted for 31% of total in 2016 – little changed from 2010
- Infrastructure investment absorbed 25% of total in 2016 – same as in 2010
- No change in share of social services which includes investment in health and education...
- ...but a significant increase in spending on environmental projects and protection

February 2017 Slide 7

C H R Metals — Concern about China's debt

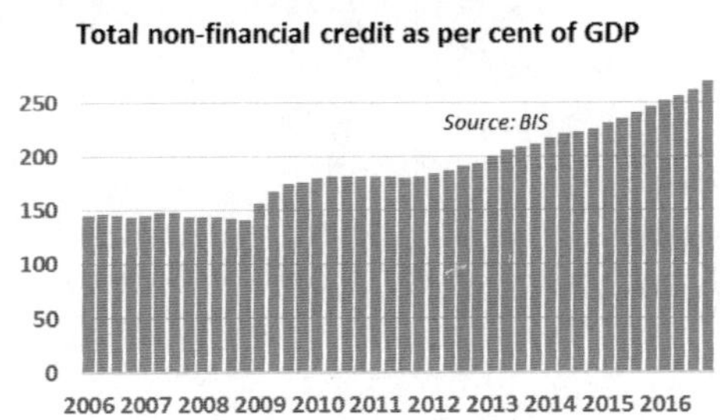

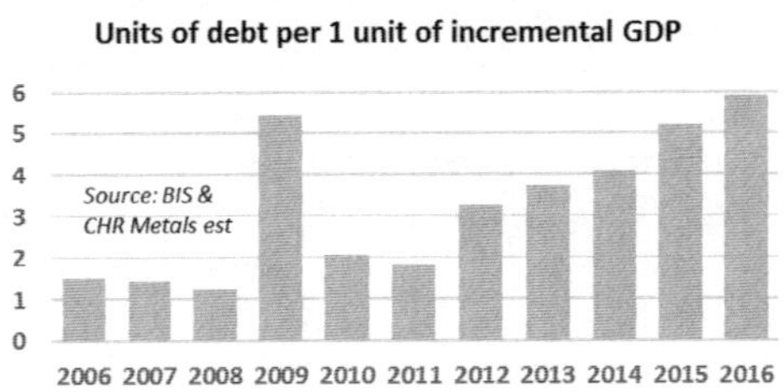

- China's debt to GDP ratio is large, but by no means unprecedented
- Many mature economies have higher ratios...
- ...but China's debt ratio is large for an emerging economy
- Most countries with high debt ratios have seen relatively slow change since 2005
- Rate of increase in China's debt ratio has outpaced most countries
- Debt becoming less efficient as a means of generating growth

February 2017 Slide 8

C H R Metals

Should we be worried?

- China's debt may be large but most is held domestically
- China not vulnerable to a sudden withdrawal of foreign capital
- Beijing will continue to underwrite losses
- Investment in infrastructure - power, roads, rail which will benefit economic development in the long run even if near-terms returns are minimal
- But, diminishing returns on investment suggest too much support for old sectors of economy which retain a crucial role in stabilising growth but at significant cost...
- ...and not enough investment in new, high-tech industries and services which must be the generators of future growth

February 2017 Slide 9

C H R Metals

Economic reform is difficult

- China's leadership is hemmed in by targets of its own making
- Plans are ambitious – a doubling of GDP per capita between 2010 and 2020, after growing 4x in previous decade...
- ...and not that far out of reach – per capita income rose 60% in five years to 2015 to reach almost US$15k (in ppp terms)
- But can a reformed and modernised economy relying more on high-tech industries, services and consumer demand deliver sustained GDP growth of 6.5% per annum (and strong demand growth for metals)?
- Inefficient state-owned enterprises survive and commitment to infrastructure spending remains a key component of current policy, despite soaring debt

February 2017 Slide 10

C H R Metals — Reforms are possible

- Commitment to improve the environment has been a key part of recent 5-year plans
- Those who have experienced choking smog of this winter in major cities might be excused for thinking little progress has been made
- Mining and smelting have been subject to intense scrutiny and there is now clear evidence that tighter regulation of these industries is happening
- Mining areas have been subject to "control" to clarify rights and licenses with specific measures ordered to improve safety and environmental performance

February 2017 Slide 11

C H R Metals — Consumer spending is expanding

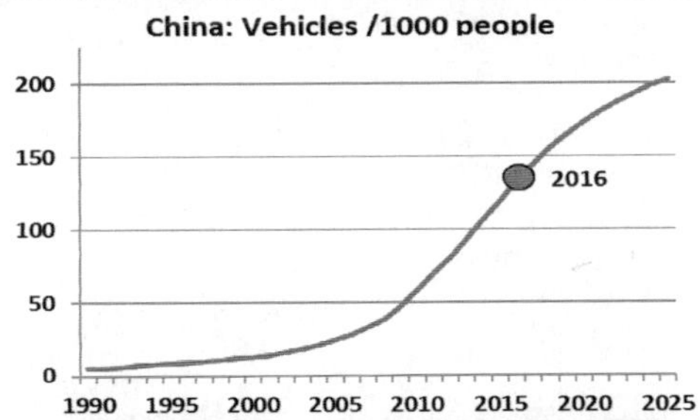

- China has a very large and expanding middle class keen to spend
- Rapid increase in auto sales
- Very large market for restaurants and entertainment
- Overseas travel is booming – more than 120M foreign visits last year
- Online shopping rose 26% in 2016 to over US$750bn – 15% of retail sales...
- ...but, shopping malls are closing even as more are being built in part to satisfy local government need for land sales revenue and GDP

February 2017 Slide 12

C H R Metals

Closing observations

- China's development over past 25 years has been very impressive, transforming the lives of several 100M people
- Growth in per capita income in last 10 years was at pace many emerging economies took 25 years or more to achieve
- China is already a highly urbanised society – many rural areas depopulated
- Reverse migration is occurring, but people are returning to new cities which have sprung up in the villages and towns they left behind
- Economic policies directing huge spending on fixed asset investment must change – misallocation of capital a significant problem
- Growth fuelled by debt is unsustainable, whoever holds that debt
- Expect to see spending on construction and physical infrastructure as a share of GDP to fall
- China will continue to be a very significant consumer of metals but economic reform will slow, or even reverse, *growth* in demand

February 2017 Slide 13

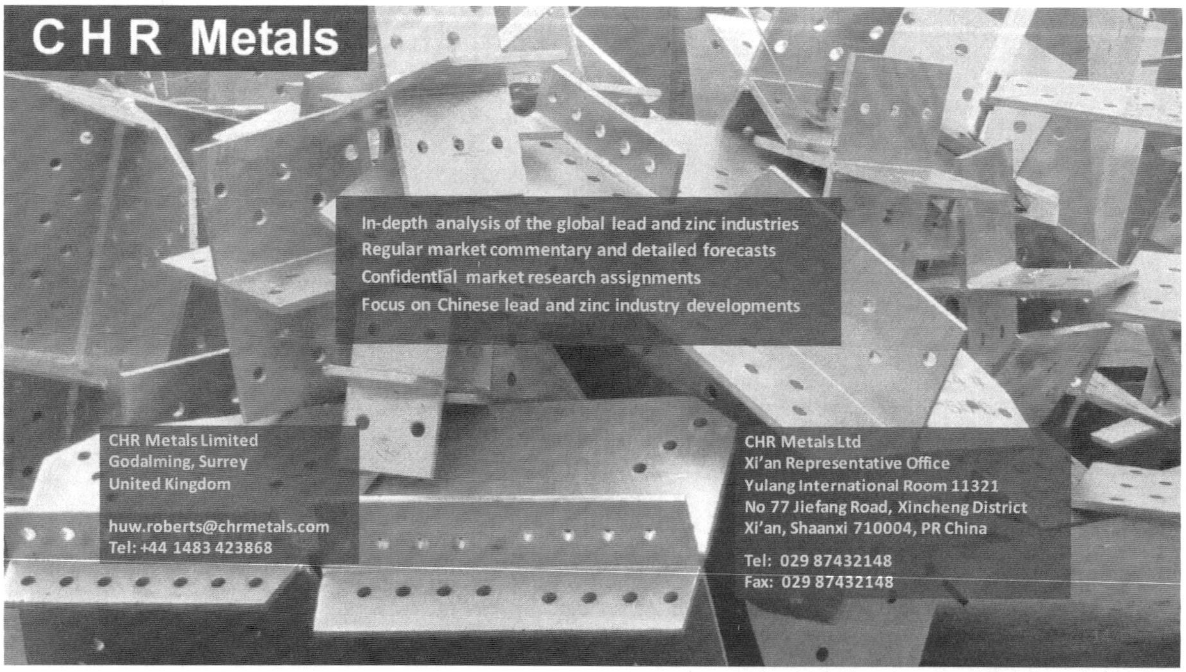

Wood Mackenzie
A Verisk Analytics Business

Zinc concentrate market outlook

IZA California
February 2017

Trusted commercial intelligence
www.woodmac.com

Agenda

1. Getting to a tight market
2. Production cuts and concentrate flows
3. Effects on treatment charges
4. Future mine supply
5. Global balance and longer-term trends

Wood Mackenzie
A Verisk Analytics Business

Sources of mine production capability (kt Zn)

Will supply from new projects be sufficient?

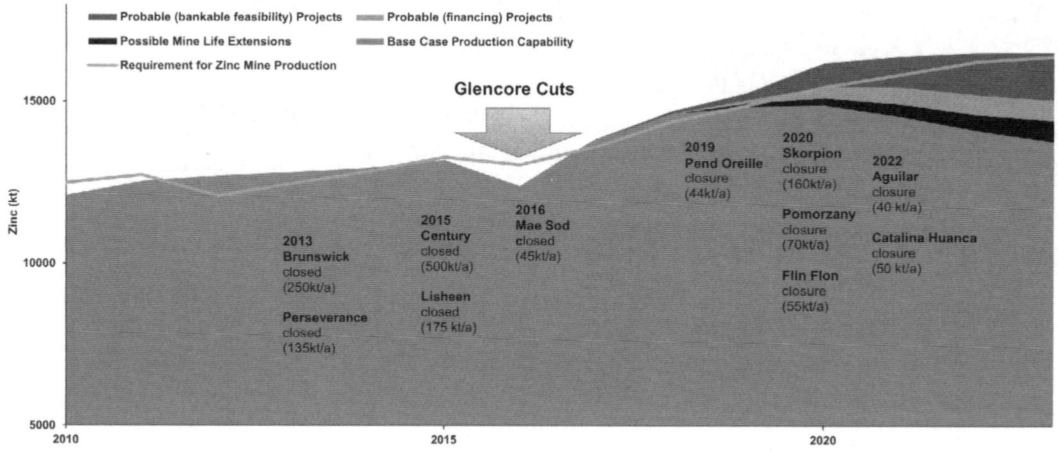

Source: Wood Mackenzie

Closures and production cuts total 1.2Mt Zn in 2016 and 1.1Mt Zn 2017...assuming a full reactivation of Glencore assets in Q4

Closures and price-induced zinc mine production capability losses (kt Zn)

Mine			2015	2016	2017
Lisheen	Ireland	Reserve depletion		105	105
Century	Australia	Reserve depletion		395	395
Mae Sod	Thailand	Reserve depletion		24	40
Total planned				**524**	**540**
Wolverine, Campo Morado, Myra Falls	North America	Early 2015	85	95	89
Middle Tennessee	USA	Late 2015	5	55	15
Al Masane	Saudi Arabia	Jan 2016 -Q4 2016		20	2
Endeavor	Australia	H2 2016		20	5
Broken Hill	Australia			20	25
Doe Run	USA			3	
Total price/technical			**90**	**213**	**136**
Mount Isa	Australia	Glencore	50	245	185
McArthur River	Australia	Glencore	25	135	100
Iscaycruz	Peru	Glencore	15	80	60
Kazzinc	Kazakhstan	Glencore	10	40	30
Total Glencore strategic			**100**	**500**	375
Grand total			**190**	**1237**	**1051**

Exports from Australia, Peru and Mexico diverted to Europe and Asia from China

Zinc and bulk concentrate exports (wet kt concentrate): 2015 minus 2016

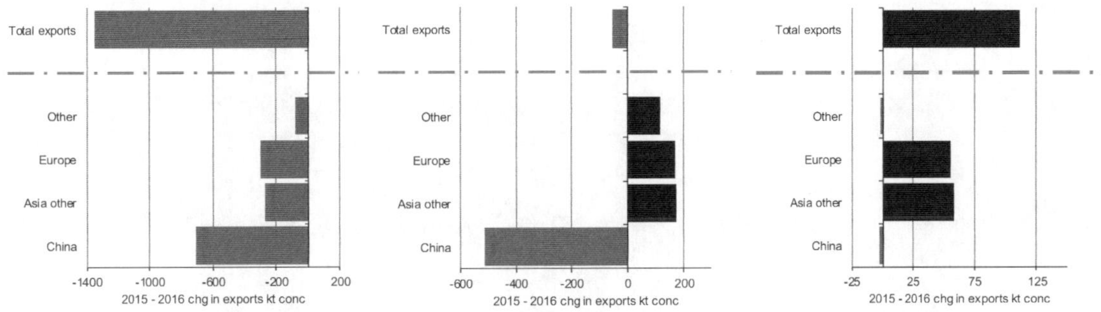

Australia
Concentrate exports
2015 2926kt
2016 1571kt

Peru
Concentrate exports
2015 1958kt
2016 1905kt

Mexico
Concentrate exports
2015 622kt
2016 734kt

Source: Wood Mackenzie, GlobalTradeTracker

Concentrate exports to China rise from other countries in 2016

Zinc and bulk concentrate exports (wet kt concentrate): 2015 minus 2016

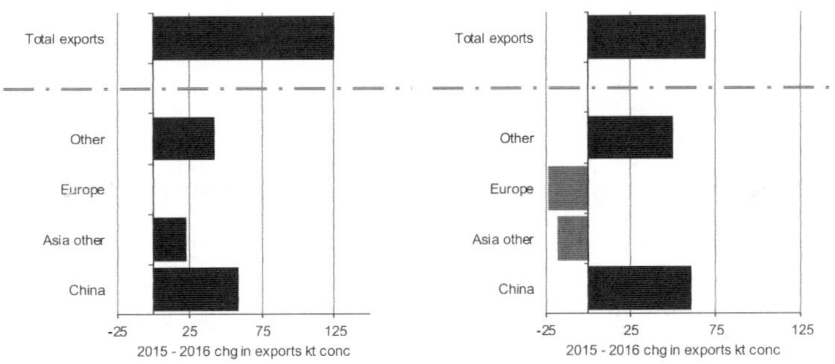

Russia
Concentrate exports
- 2015 116kt
- 2016 241kt

Bolivia
Concentrate exports
- 2015 875kt
- 2016 944kt

Chinese imports increased from 14 countries totalling over 250kt conc

North Korea	74kt
Eritrea	46kt
Iran	39kt
Mongolia	34kt
Myanmar	11kt
USA	11kt
Pakistan	8kt

Source: Wood Mackenzie, GlobalTradeTracker

Chinese concentrate imports plummeted in 2016

Chinese sources of concentrate imports

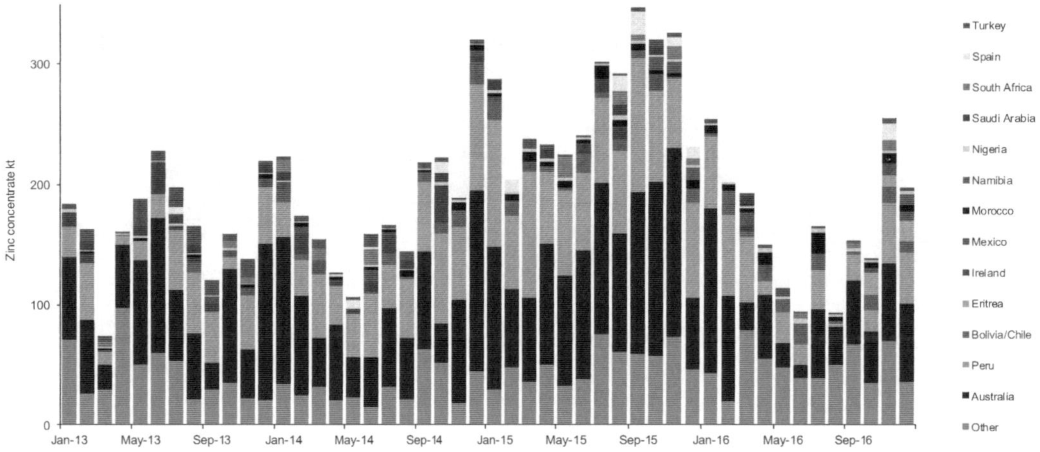

Source Customs Authorities, Wood Mackenzie

Chinese and rest of world concentrate markets have moved to deficit

Rest of World concentrate balance | **Chinese concentrate balance**

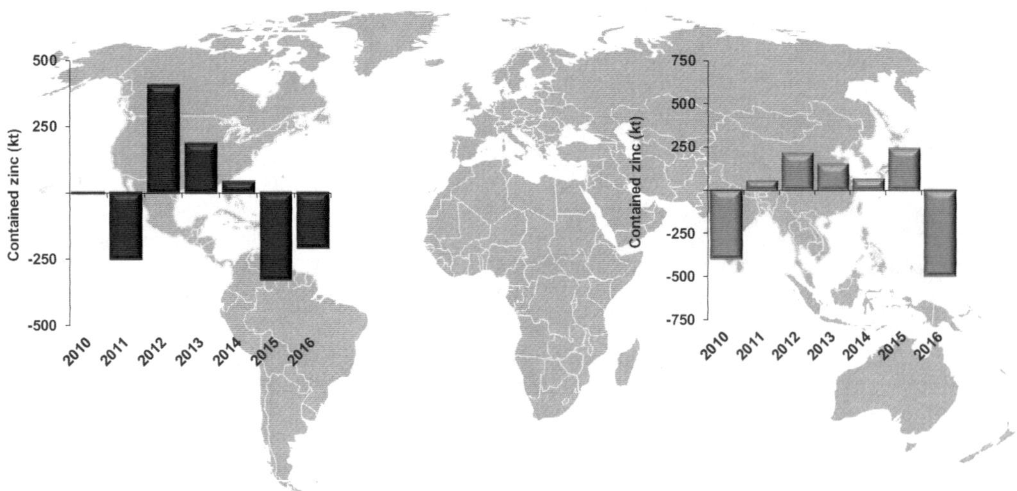

The deficits have depleted concentrate stocks and forced treatment charges lower

Lower spot TC induce ROW miners to export to China

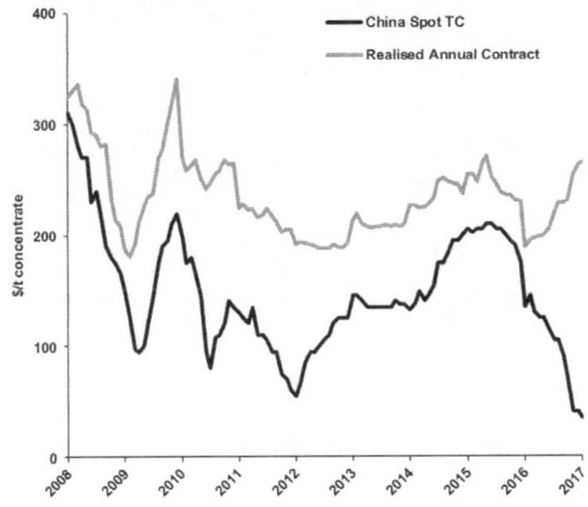

Source: Wood Mackenzie

Chinese smelter's incentive to buy imported conc

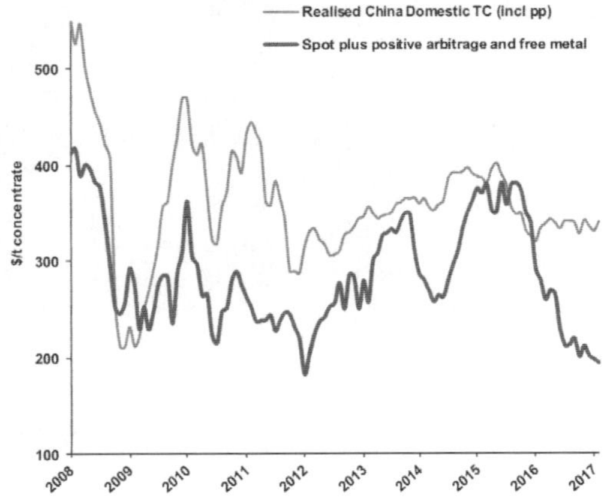

Source Wood Mackenzie

Historical zinc mine price elasticity of supply

Annual change in zinc mine output vs price

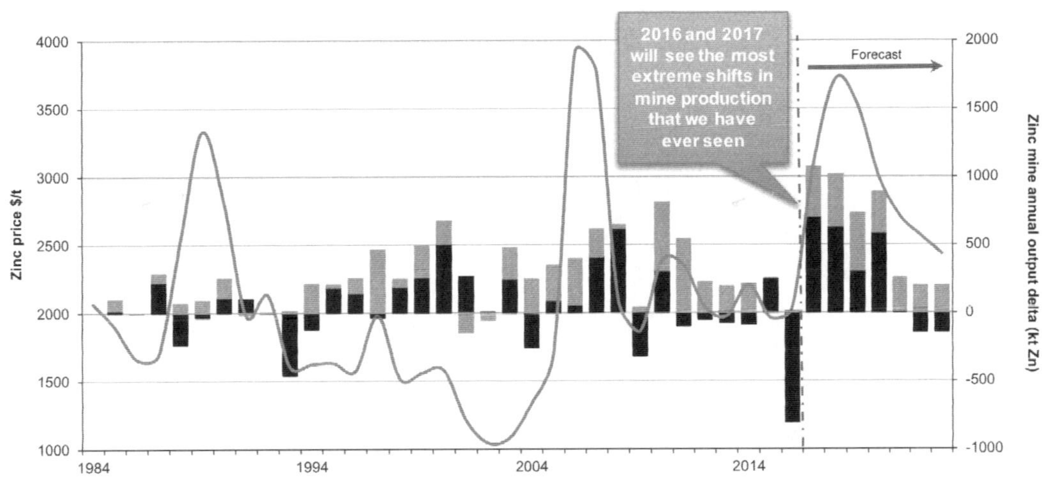

Source: Wood Mackenzie

Project development and high/low operating cost split *vs* cycle average zinc price

Phases of historical project development

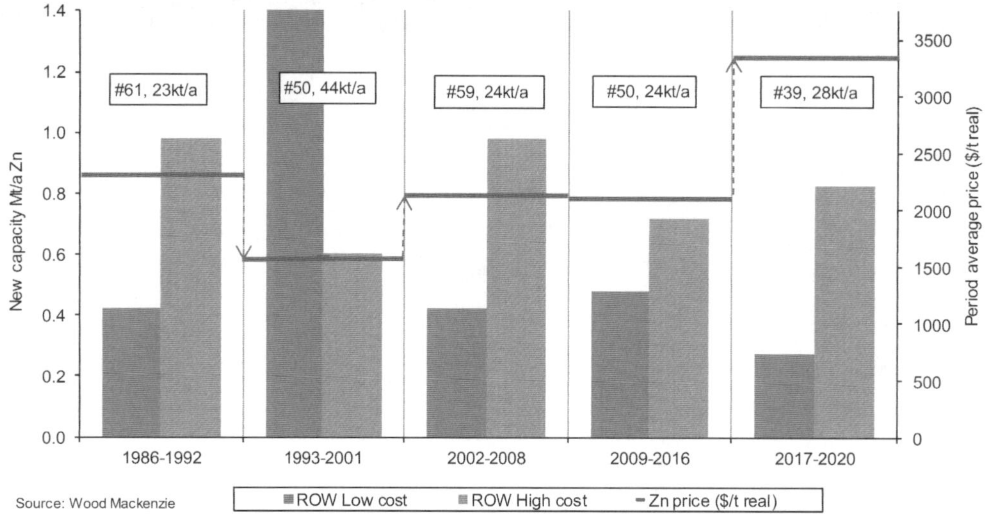

Source: Wood Mackenzie

Wood Mackenzie
A Verisk Analytics Business

Base case and probable projects have potential to generate 2.3Mt/a of new capacity by 2020

Population of operating mines, base case projects and probable projects

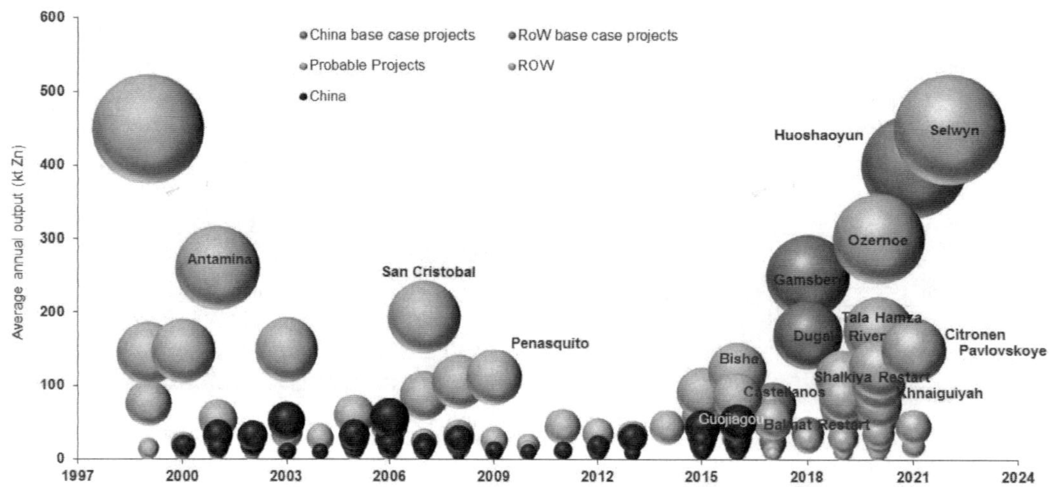

Wood Mackenzie
A Verisk Analytics Business

Once permitted and financed zinc mines can be built quickly

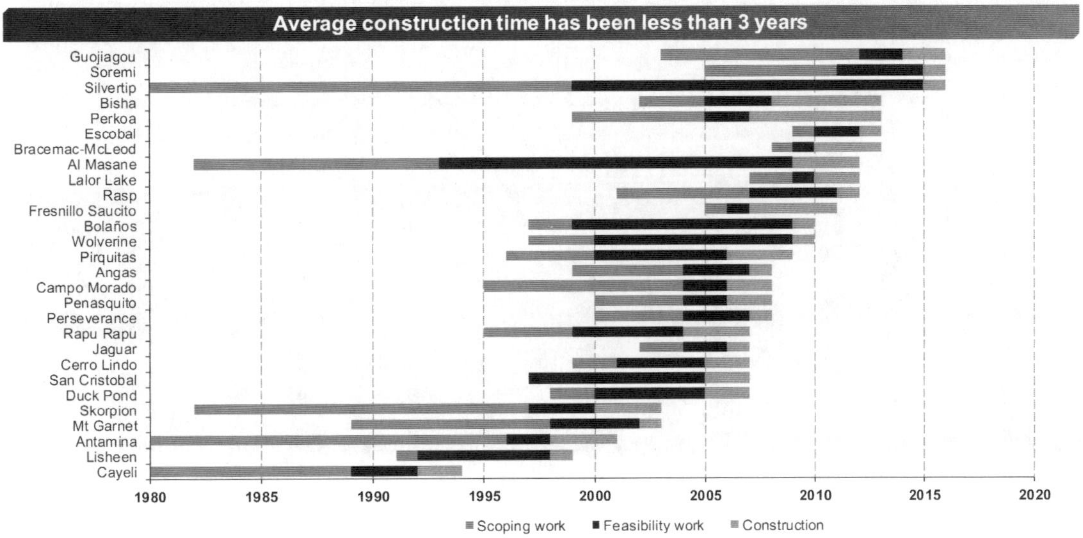

Source: Wood Mackenzie

Smelter output constrained by tight concentrate market resulting in a draw-down of refined stocks

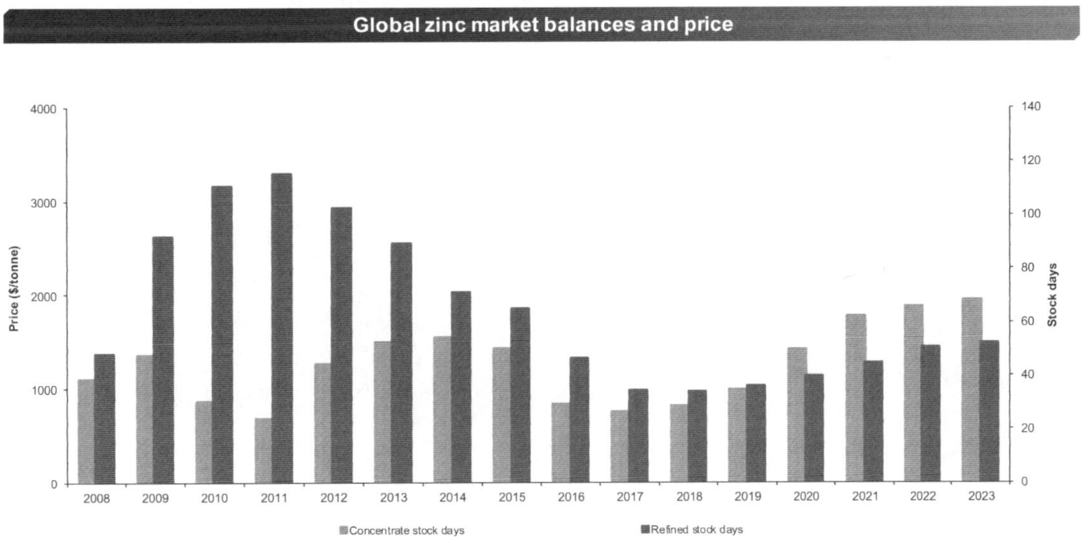

Smelter output constrained by tight concentrate market resulting in a draw-down of refined stocks. Prices to peak in 2018

Global zinc market balances and price

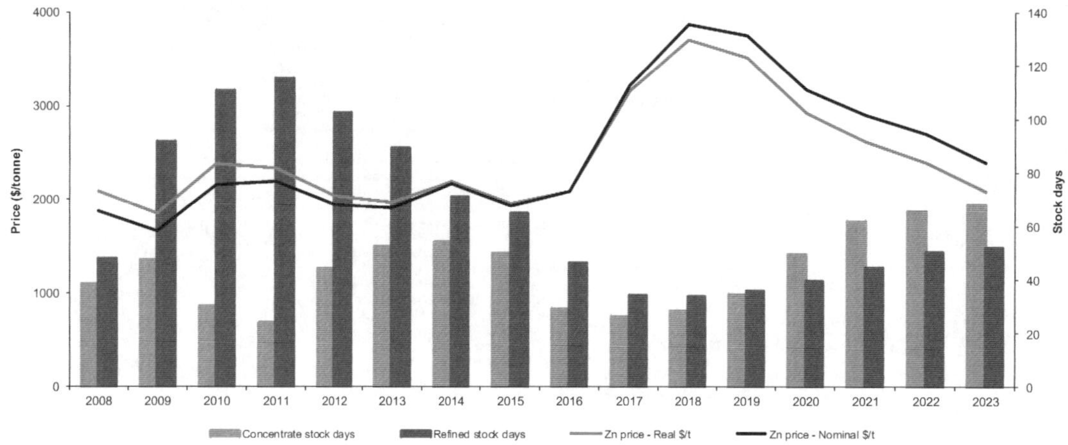

Wood Mackenzie
A Verisk Analytics Business

Possible threats to the current tight concentrate market

Glencore Reactivations

- **Rapid reactivation of Glencore mines**
  - Why kill the price rally, destroying the return on investment in mine cuts?
  - Glencore has said that reactivation "might be staged over a period of time as a way of testing the market", with "the option of bringing them back one by one rather than all together"

China Mines

- **Double-digit growth in China mine output**
  - Occurred in 6 of the past 15 years
  - We assume strong price induced growth in Chinese mine production to 2020
  - BUT, growth will be constrained by environmental regulation

ROW Mines

- **Strong growth in ROW mine output**
  - New projects are on the way but have been small so far
  - Larger projects are seeing development activity, but awaiting finance

Wood Mackenzie
A Verisk Analytics Business

Conclusions

- Glencore's mine production cuts have transformed the outlook
- Tight concentrate market emerged in Q3 2016
- Normal trade flows of concentrate are being disrupted
- Tight refined market emerges around mid-2017
- Timing of the Glencore's restarts is a critical risk
- New mine development has been slow. Uncertainty remains over new project development, especially the larger probable projects
- Lagged development of new capacity will keep market fundamentals positive

Disclaimer

Wood Mackenzie Disclaimer

- This presentation has been prepared by Wood Mackenzie. Its contents and conclusions are confidential and may not be disclosed to any other persons or companies without Wood Mackenzie's prior written permission.
- The information upon which this report is based comes from our own experience, knowledge and databases. The opinions expressed in this report are those of Wood Mackenzie. They have been arrived at following careful consideration and enquiry but we do not guarantee their fairness, completeness or accuracy. The opinions, as of this date, are subject to change. We do not accept any liability for your reliance upon them.

Wood Mackenzie
A Verisk Analytics Business

Jonathan Leng

Principal Analyst – Zinc Markets

T +44 (0)1483 596 626
E jonathan.leng@woodmac.com

- Jonathan is a principal analyst focused on zinc mine and smelter supply and concentrate analysis, contributing to the Wood Mackenzie zinc market outlook
- Prior to his current role, Jonathan was manager of the zinc, lead and gold mine cost research teams
- He has over 18 years of metals and mining experience, primarily in base metal, ferroalloy and gold research
- Prior to joining Wood Mackenzie, he worked as a geologist for Croesus Gold in Australia and in mining and metals research at CRU. Jonathan also worked in construction cost management at Davis Langdon
- Jonathan holds a M.Sc. Mining and Mineral Resource Evaluation, B.Sc. (hons) Geology, and PGDip in Construction Cost Management

Wood Mackenzie
A Verisk Analytics Business

Europe +44 131 243 4400
Americas +1 713 470 1600
Asia Pacific +65 6518 0800

Email contactus@woodmac.com
Website www.woodmac.com

INVESTORS IN PEOPLE | Gold

Christopher Parker
Research Director Zinc Markets
T +44 (0) 148 359 6616
E chris.parker@woodmac.com

Andrew Thomas
Principal Analyst
T +44 (0) 148 359 6611
E andrew.thomas@woodmac.com

Jonathan Leng
Principal Analyst
T +44 (0) 148 359 6626
E jonathan.leng@woodmac.com

Wood Mackenzie
A Verisk Analytics Business

CRU THE INDEPENDENT AUTHORITY MINING | METALS | FERTILIZERS

1

There is a clear justification for even stronger zinc prices

What we know

- Zinc stocks grew sharply during the great recession
  - Some of this was concentrate; some was metal
  - Producers cut back, but consumption suffered more
- Mine cutbacks and a Chinese slowdown have now reversed the flow

What we don't know

- How stocks moved in the intervening years
  - What was the balance between mine and refined production ?
  - What was the balance between refined production and consumption ?
  - What was the balance between mine production and metal offtake ?
- And producers know just as little as us in China and the West

What this means

- The market will be reactive rather than proactive going forward
- Supply may not grow before prices are very high, leaving the market to be re-balanced by changing demand

www.crugroup.com

CRU THE INDEPENDENT AUTHORITY MINING | METALS | FERTILIZERS

2

Stocks accumulated in 2008-09 are now being drawn down

global developments; million tonnes recoverable zinc

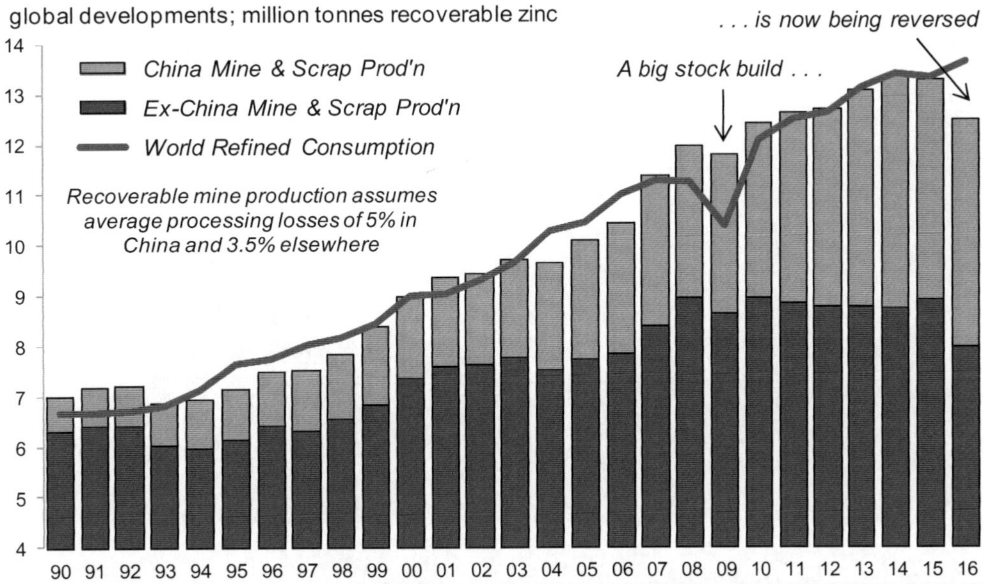

Source: January 2017 Zinc Market Outlook

www.crugroup.com

THE INDEPENDENT AUTHORITY
MINING | METALS | FERTILIZERS

All excess conc. stocks may have been used by early 2017

cumulative concentrate stock surplus; adjusted; million tonnes zinc

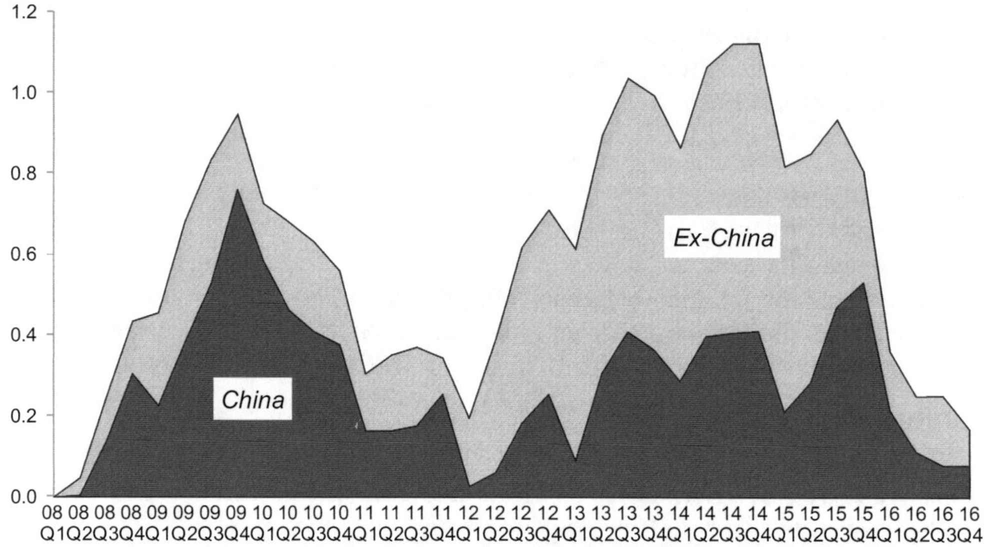

Source: January 2017 Zinc Market Outlook

www.crugroup.com

THE INDEPENDENT AUTHORITY
MINING | METALS | FERTILIZERS

Spot TCs reflect this, especially for Chinese imports

US$/t concentrate

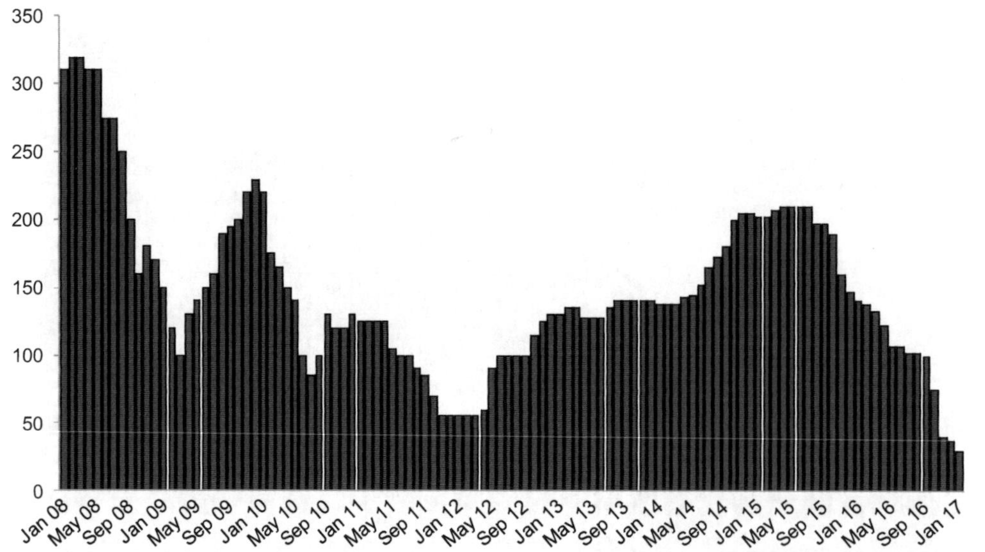

Source: January 2017 Zinc Market Outlook

www.crugroup.com

THE INDEPENDENT AUTHORITY
MINING | METALS | FERTILIZERS

5

However, the fall in LME metal slowed in 2015-16

million tonnes

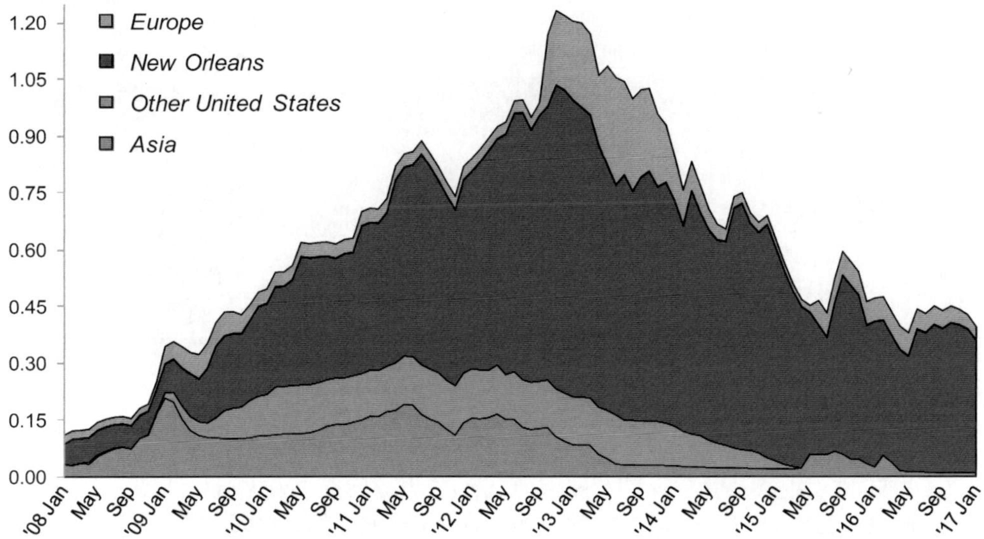

Source: January 2017 Zinc Market Outlook www.crugroup.com

THE INDEPENDENT AUTHORITY
MINING | METALS | FERTILIZERS

6

Total metal stocks may still be excessive

cumulative refined stock surplus; adjusted; million tonnes

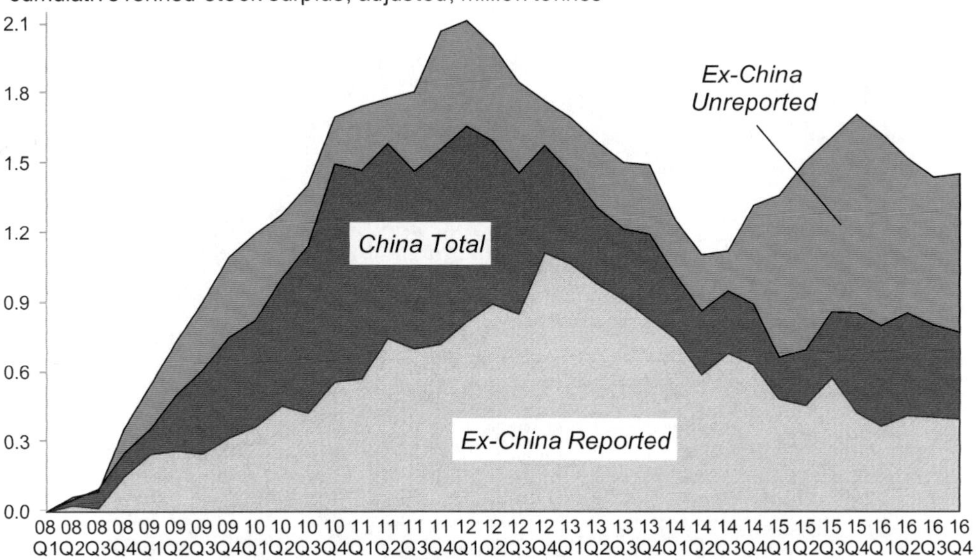

Source: January 2017 Zinc Market Outlook www.crugroup.com

CRU THE INDEPENDENT AUTHORITY MINING | METALS | FERTILIZERS

7

Spot premia have fallen from a 2013-14 high

regional premia for SHG; US$/t

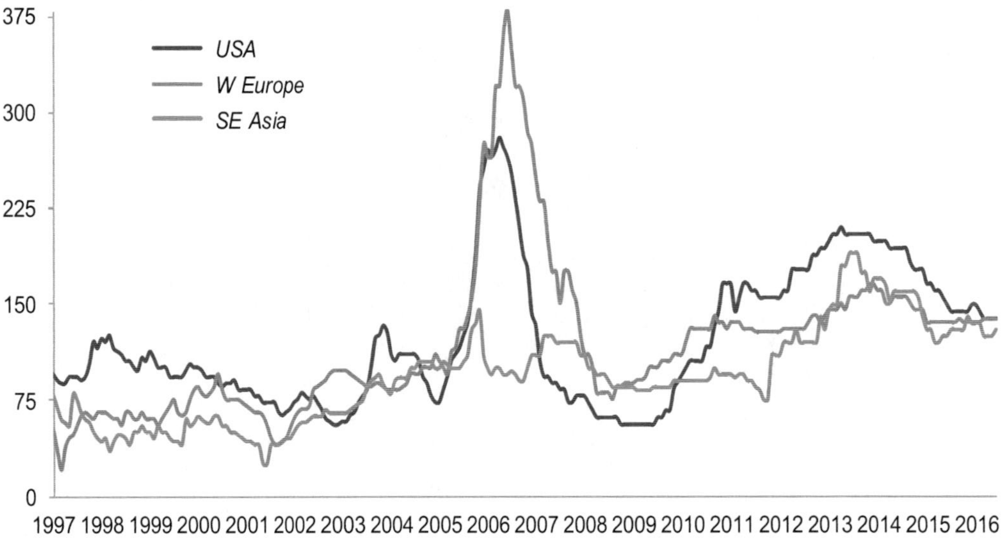

Source: January 2017 Zinc Market Outlook

www.crugroup.com

CRU THE INDEPENDENT AUTHORITY MINING | METALS | FERTILIZERS

8

Resurgent investor demand may have been needed to push last year's prices up

2007	2008	2009	2010	2011	2012	2013	2014	2015	2016
EME 37.5%	Govt bonds 52.6%	EME 59.4%	REITS 31.6%	EMD 10.0%	REITS 14.9%	DM Equities 25.0%	REITS 35.1%	REITS 6.2%	HY bonds 36.3%
Cmdty 14.7%	IG bonds 26.5%	HY bonds 41.9%	EME 22.9%	REITS 6.1%	HY bonds 14.4%	Hedge Funds 9.6%	EMD 12.8%	EMD 7.7%	Cmdty 32.9%
Hedge Funds 12.6%	EMD 25.0%	Hedge Funds 18.4%	Cmdty 20.5%	Govt bonds 7.1%	EME 13.4%	Portfolio 6.0%	DM Equities 12.1%	DM Equities 5.5%	EMD 30.8%
Portfolio 8.9%	Cash 6.9%	Portfolio 16.5%	HY bonds 18.4%	IG bonds 5.1%	EMD 12.9%	HY bonds 5.3%	IG bonds 9.6%	HY bonds 2.9%	REITS 30.4%
Govt bonds 8.7%	Portfolio 1.3%	DM Equities 16.4%	DM Equities 15.9%	HY bonds 3.9%	DM Equities 11.4%	REITS 1.3%	Portfolio 8.7%	Govt bonds 2.3%	EME 29.5%
DM Equities 7.7%	HY bonds 1.2%	REITS 13.5%	EMD 15.3%	Cash 1.2%	Portfolio 8.0%	Cash 0.5%	HY bonds 6.2%	IG bonds 2.0%	DM Equities 25.6%
Cash 6.1%	Cmdty -12.5%	EMD 12.1%	Portfolio 14.9%	Portfolio -0.9%	Hedge Funds 7.5%	IG bonds -1.5%	Govt bonds 5.4%	Portfolio 1.3%	IG bonds 24.4%
IG bonds 4.9%	REITS -13.2%	Cmdty 7.3%	Hedge Funds 10.6%	Hedge Funds -2.9%	IG bonds 6.3%	EME -4.1%	EME 4.3%	Cash 0.7%	Portfolio 24.4%
EMD 4.7%	DM Equities -17.4%	IG bonds 6.1%	Govt bonds 3.2%	DM Equities -4.3%	Cash 1.4%	Govt bonds -6.1%	Hedge Funds 4.1%	Hedge Funds -0.0%	Govt bonds 21.3%
HY bonds 1.4%	Hedge Funds -18.3%	Cash 2.2%	IG bonds 9.2%	Cmdty -12.7%	Govt bonds -2.6%	EMD -10.0%	Cash 0.6%	EME -0.7%	Cash 0.7%
REITS -19.2%	EME -35.2%	Govt bonds -8.6%	Cash 1.0%	EME -17.6%	Cmdty -5.4%	Cmdty -11.2%	Cmdty -11.8%	Cmdty -20.3%	Hedge Funds -1.1%

Source: JP Morgan Asset Management

www.crugroup.com

The 2016 market was in large deficit overall

global market balances; '000 tonnes contained zinc

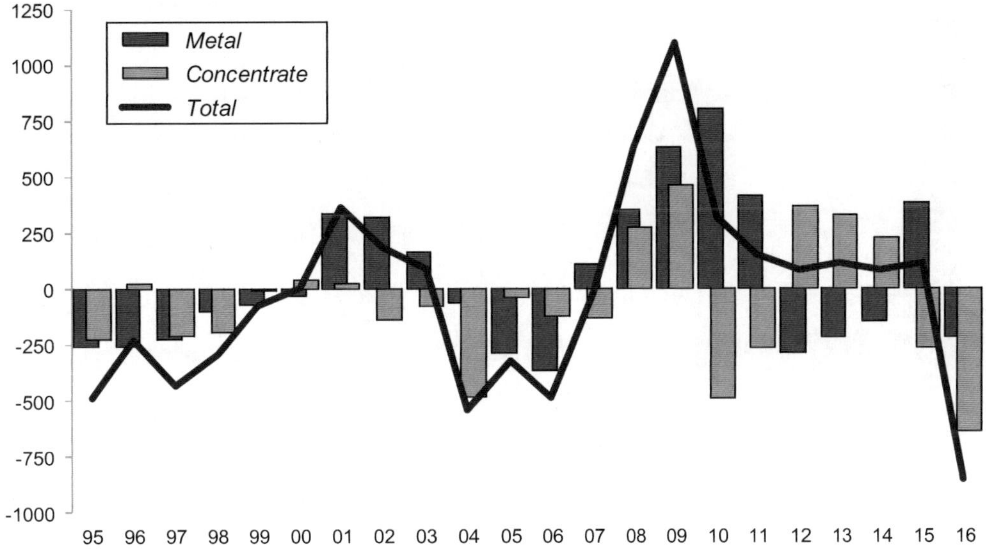

Source: January 2017 Zinc Market Outlook

www.crugroup.com

But this took major mine closures to achieve

2015-16 production change; '000t zinc

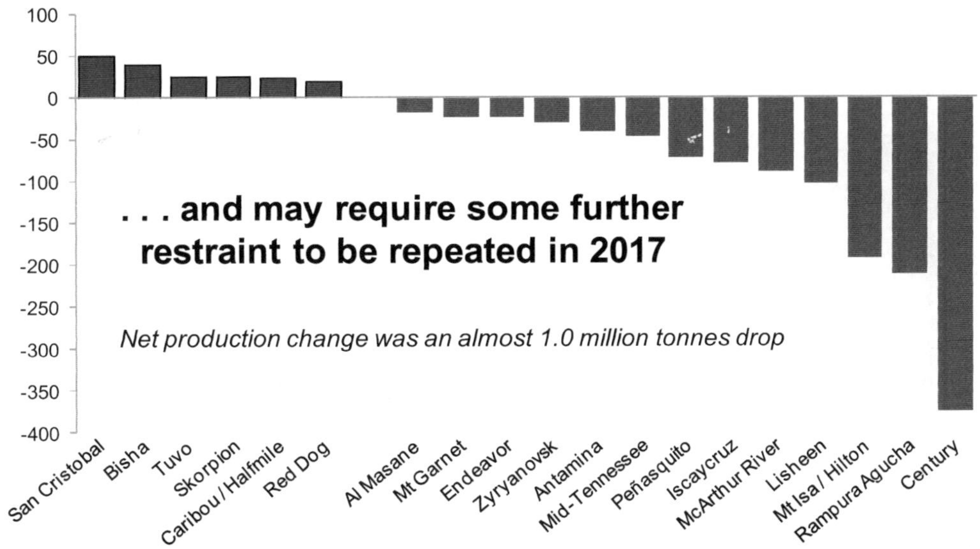

Source: January 2017 Zinc Market Outlook

www.crugroup.com

Numerous uncommitted mines could in fact be developed in the West

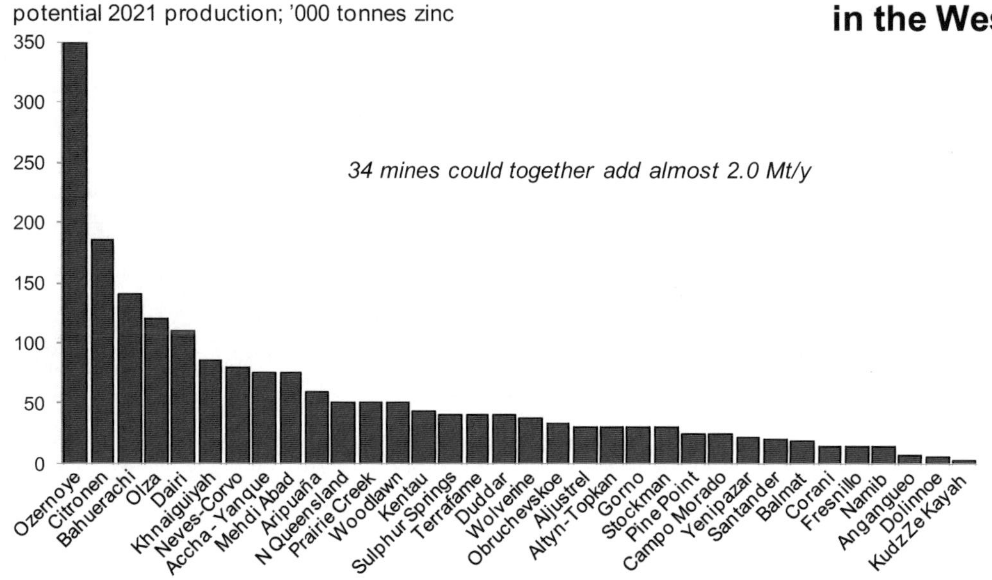

Source: January 2017 Zinc Market Outlook

2016-21 growth may be significant in China too

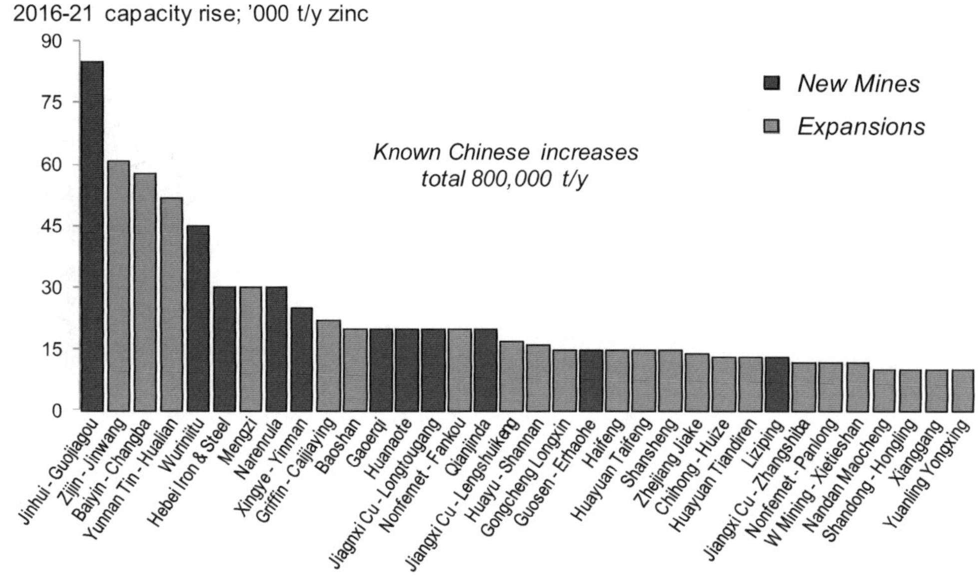

Source: January 2017 Zinc Market Outlook

We even assume that Chinese mine production growth will soon repeat past highs

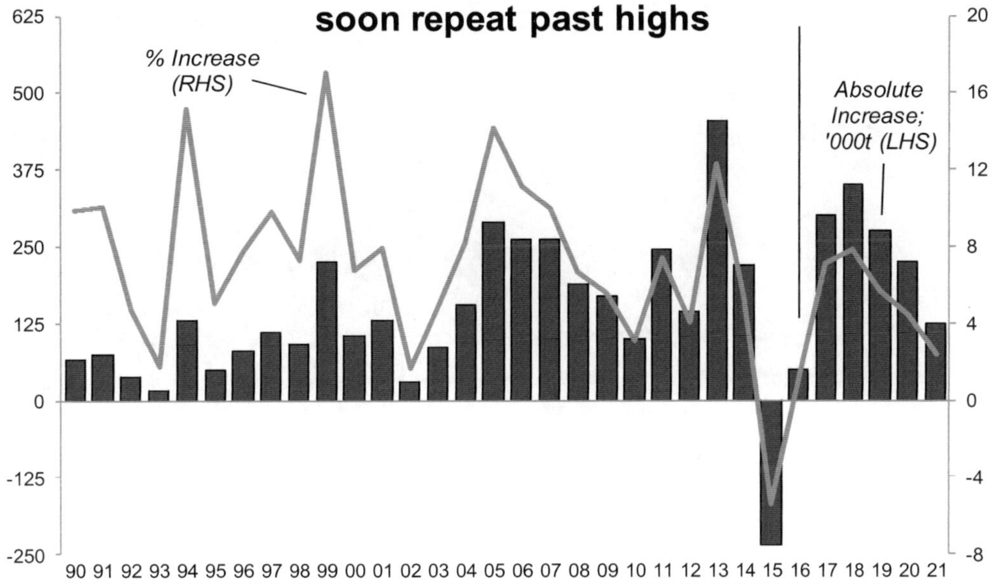

Source: January 2017 Zinc Market Outlook www.crugroup.com

However, Western reactivations may be more restrained

2016-17 production change; '000t zinc

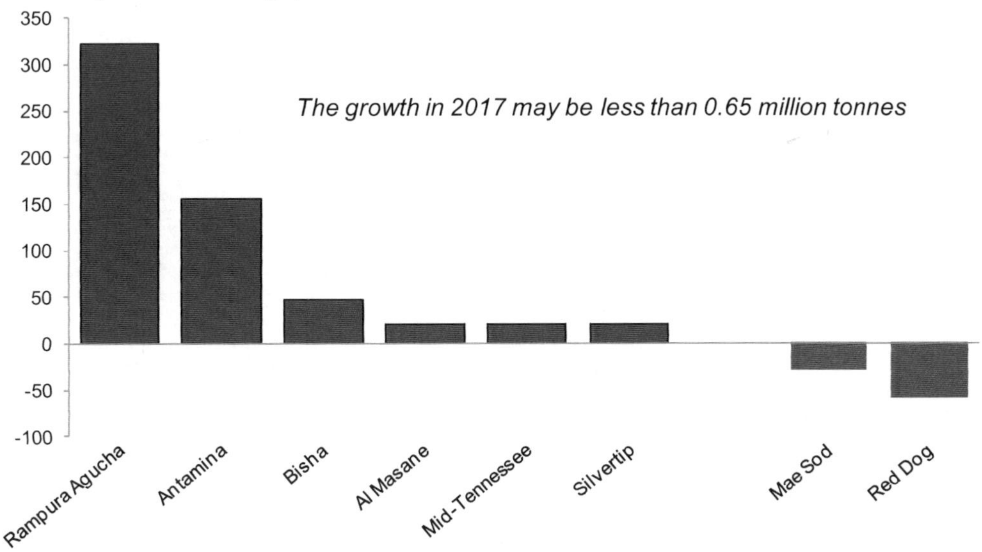

Source: January 2017 Zinc Market Outlook www.crugroup.com

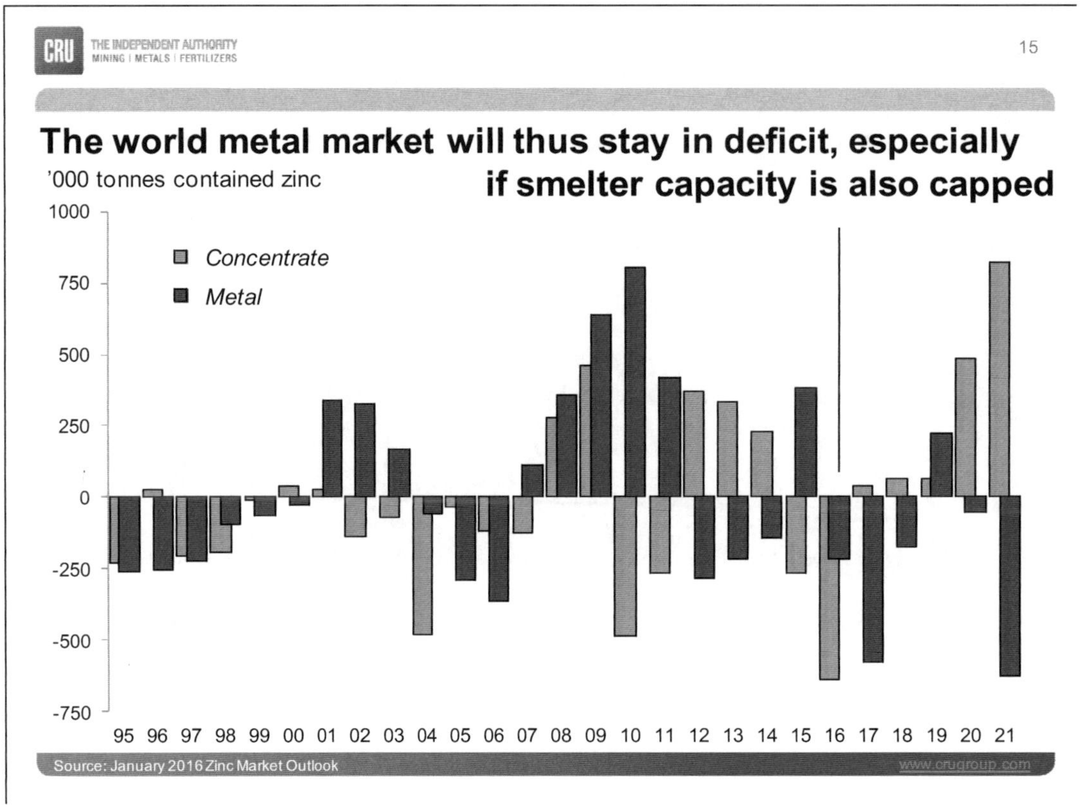
CRU THE INDEPENDENT AUTHORITY
MINING | METALS | FERTILIZERS
15
The world metal market will thus stay in deficit, especially if smelter capacity is also capped
'000 tonnes contained zinc
Concentrate
Metal
1000
750
500
250
0
-250
-500
-750
95 96 97 98 99 00 01 02 03 04 05 06 07 08 09 10 11 12 13 14 15 16 17 18 19 20 21
Source: January 2016 Zinc Market Outlook
www.crugroup.com

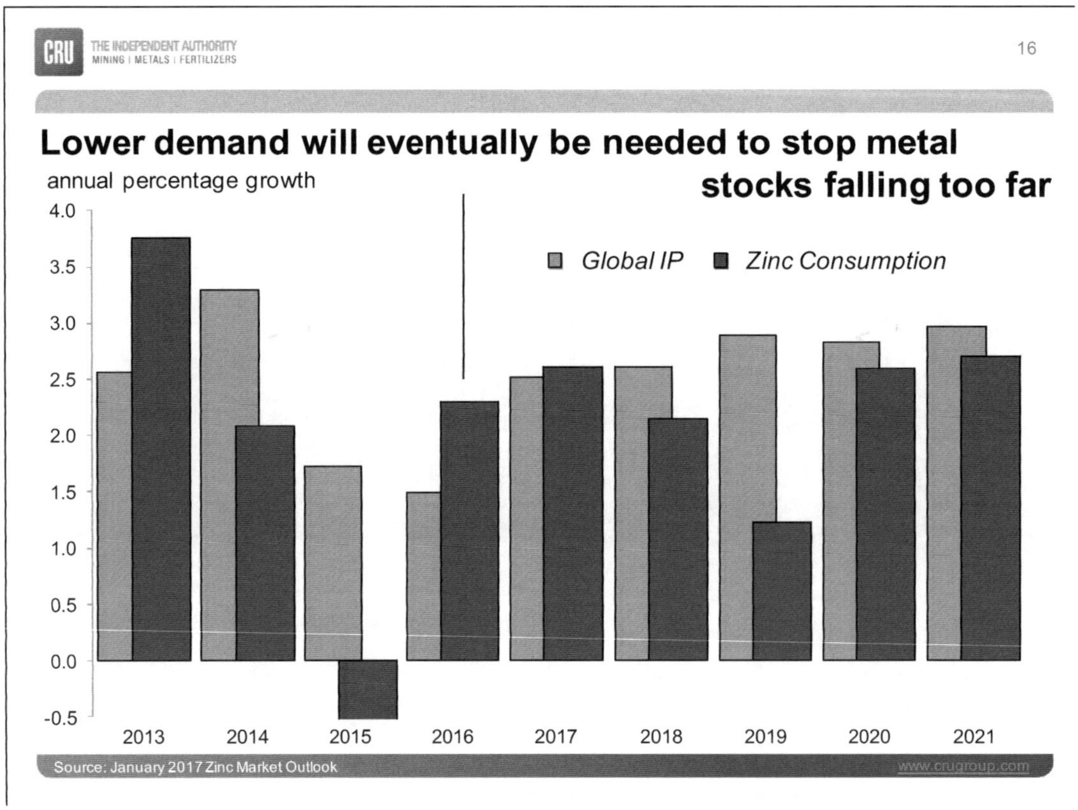
CRU THE INDEPENDENT AUTHORITY
MINING | METALS | FERTILIZERS
16
Lower demand will eventually be needed to stop metal stocks falling too far
annual percentage growth
Global IP
Zinc Consumption
4.0
3.5
3.0
2.5
2.0
1.5
1.0
0.5
0.0
-0.5
2013
2014
2015
2016
2017
2018
2019
2020
2021
Source: January 2017 Zinc Market Outlook
www.crugroup.com

THE INDEPENDENT AUTHORITY
MINING | METALS | FERTILIZERS

17

This may require higher prices than for any supply-side response

- **Diecasters could suffer (especially if aluminium prices are still low)**
- **Chinese efficiencies of use could improve (how many first- and second-generation plants are left?)**
- **The big threat is in automotive body sheet (this will also be permanent but will need a lot of aluminium)**

Experience shows that any of these may take more than just a price spike

www.crugroup.com

THE INDEPENDENT AUTHORITY
MINING | METALS | FERTILIZERS

18

Latest forecast has a spike in 2019 and recovery in 2021

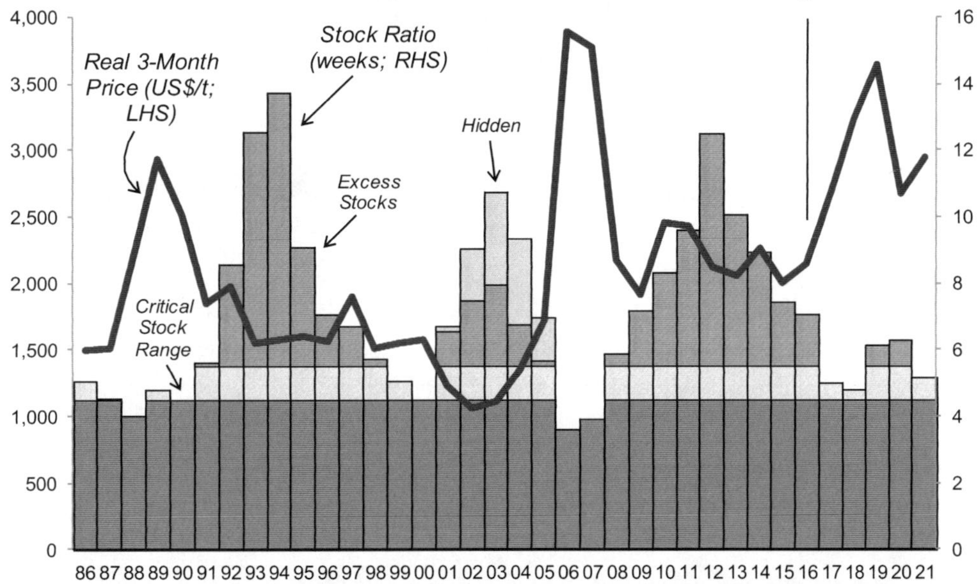

Source: January 2017 Zinc Market Outlook

www.crugroup.com

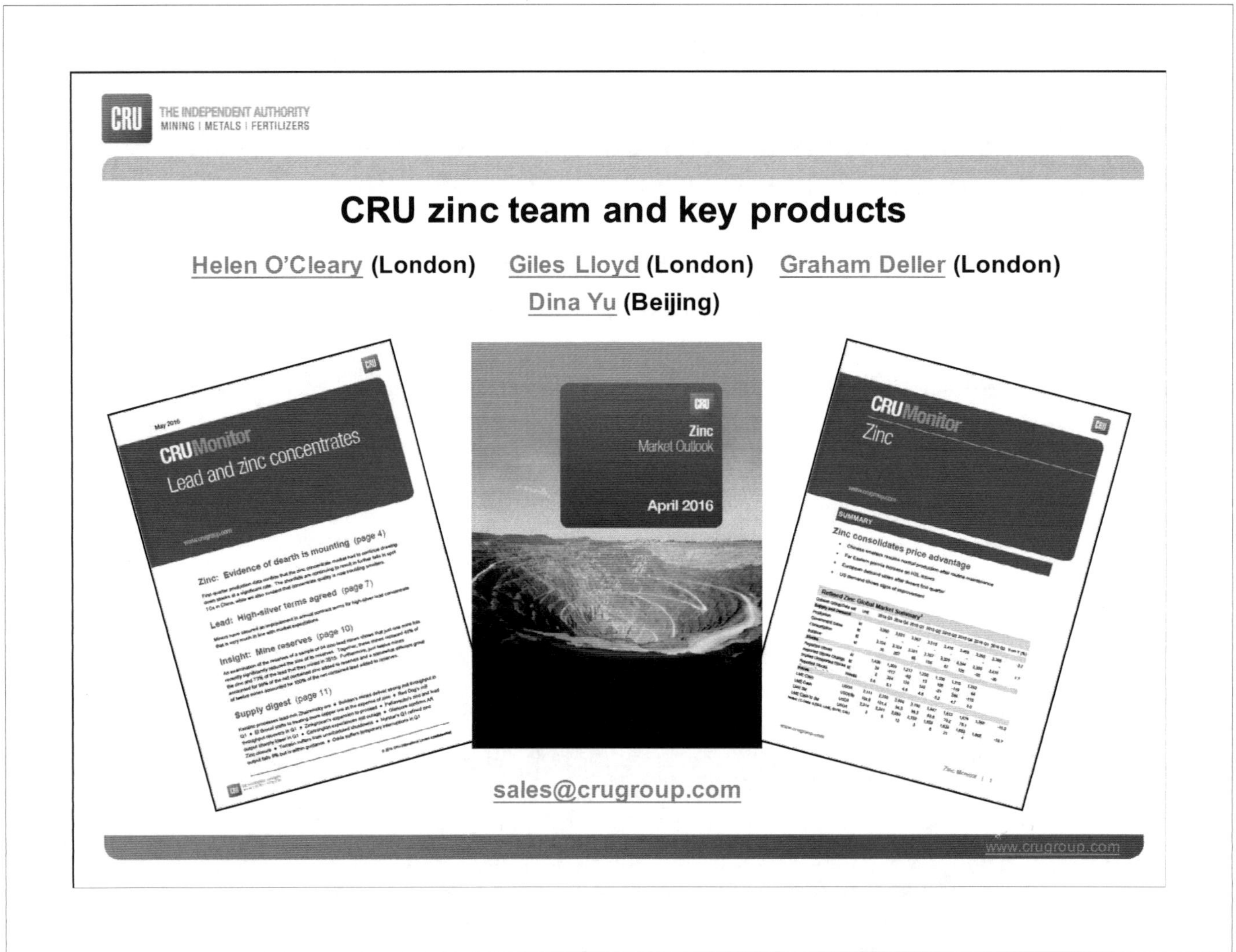
THE INDEPENDENT AUTHORITY
MINING | METALS | FERTILIZERS
CRU zinc team and key products
Helen O'Cleary (London) Giles Lloyd (London) Graham Deller (London)
Dina Yu (Beijing)
CRU Monitor
Lead and zinc concentrates
Zinc
Market Outlook
April 2016
CRU Monitor
Zinc
sales@crugroup.com
www.crugroup.com

Macquarie Research | Commodities

MACQUARIE

Zinc - Entering Phase II

Zinc market outlook

Macquarie Commodities Research
Macquarie Capital (Europe) Ltd
28 Ropemaker Street
London EC2Y 9HD, UK

Vivienne Lloyd
+44 20 3037 4530
vivienne.lloyd@macquarie.com

February 2017

Please read Disclaimer on Pages 19-21

Macquarie Research | Commodities

MACQUARIE

Macquarie Research key commodity calls

Outperformers on a 12- to 24-month basis from current levels

- → **US Natural Gas** – Strong underlying demand will extend period of tightness.
- → **Zinc** – The raw material constraint story becomes ever clearer.
- → **Silver** – Not hurt by reflation, helped by inflation, and lacking mine supply.
- → **Cobalt** – A consistent deficit expected, and major country risk remains.
- → **Uranium** – Prices are finally invoking a supply response.

Commodities for which the long-term challenges are underappreciated

- → **Potash** – The market share battle is maturing, but overcapacity persists.
- → **Alumina** – Perhaps the only commodity with significant ex-China capacity still coming.
- → **Aluminium** – Industry cost deflation is an ongoing headwind.
- → **Steel** – Chinese supply cuts will help, but a downward demand trend is the main issue.
- → **Nitrogen** – Increased efficiency of use and displacement by alternatives present problems.

Source: Macquarie Research, February 2017

Macquarie Research | Commodities

MACQUARIE

Positions of key mined/extracted commodities in the fundamental cycle*

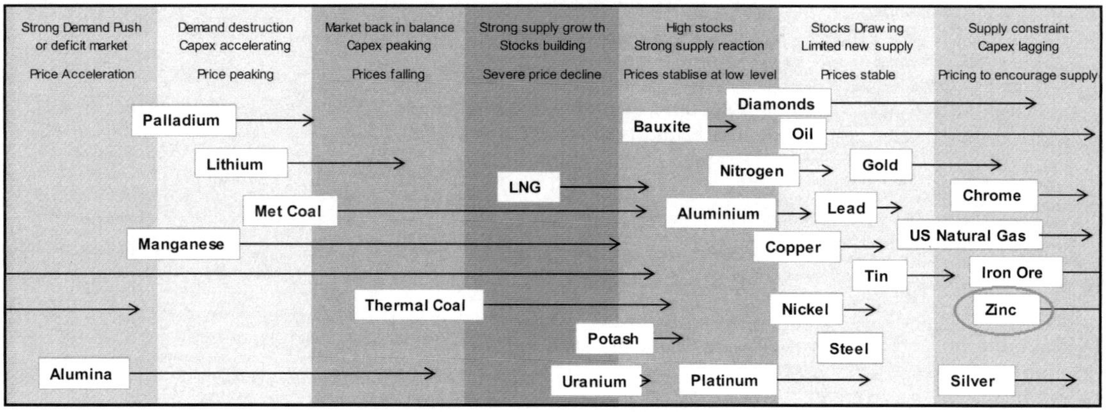

*arrow shows 2 year progression

Source: Macquarie Research, February 2017

Macquarie Research | Commodities

MACQUARIE

Spoiler alert! Zinc is in a structural deficit

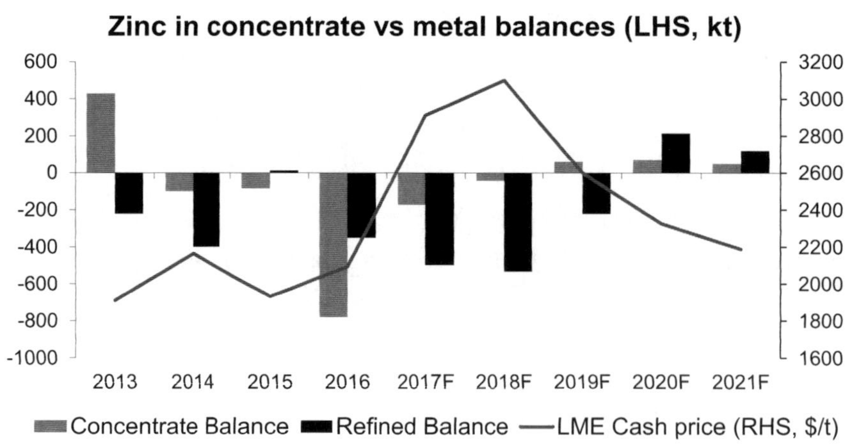

Source: Wood Mackenzie, CRU, Antaike, BGR Macquarie Research, February 2017

Macquarie Research | Commodities

MACQUARIE

Century, Lisheen, Glencore, Vedanta – it's a familiar tale

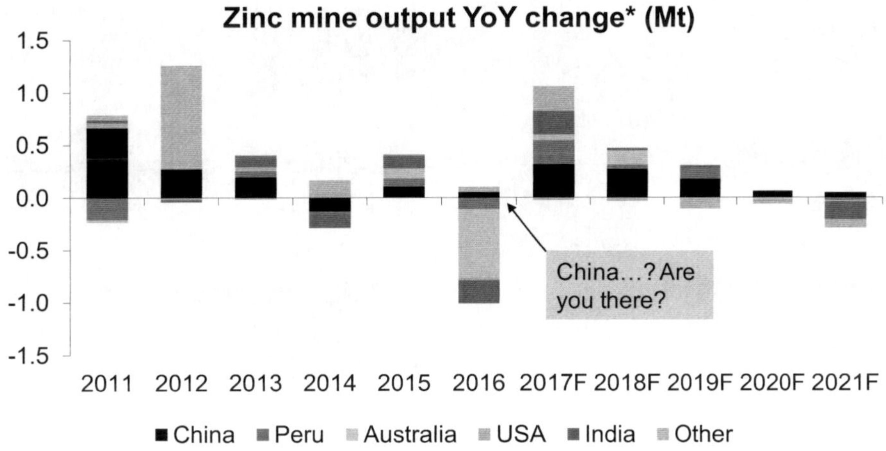

*forecast production before disruption

Source: Wood Mackenzie, CRU, Antaike, BGR Macquarie Research, February 2017

Page 5

Macquarie Research | Commodities

MACQUARIE

Changing landscape of Chinese mine output

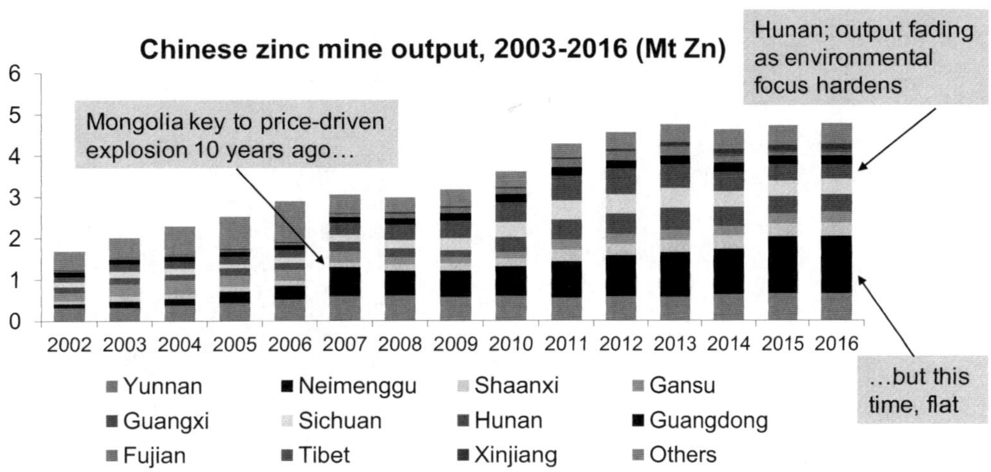

Source: Wood Mackenzie, CRU, Antaike, BGRIMM, Macquarie Research, February 2017

Page 6

Macquarie Research | Commodities

MACQUARIE

What next? A few projects coming in to address the deficit

Gamsberg (250ktpa)

Biliutai (41ktpa)

Tala Hamza (175ktpa)

Jinshi (32ktpa)

Dugald River (170ktpa)

Wolitu (30ktpa)

Guojiagou (123ktpa)

Bachagou (25ktpa)

Huangutun (20ktpa)

Bangzhong (41ktpa)

Khnaiguiyah (90ktpa)

Yaojialing (35ktpa)

...but too late to avoid refined cutbacks

Source: Wood Mackenzie, CRU, Antaike, BGRIMM, Macquarie Research, February 2017

Page 7

Macquarie Research | Commodities

MACQUARIE

Entering Phase II: The metal deficit

Concs short ~700kt

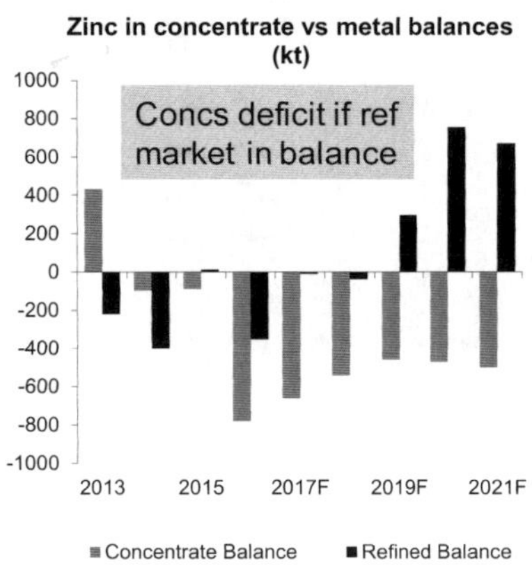

Who loses out?

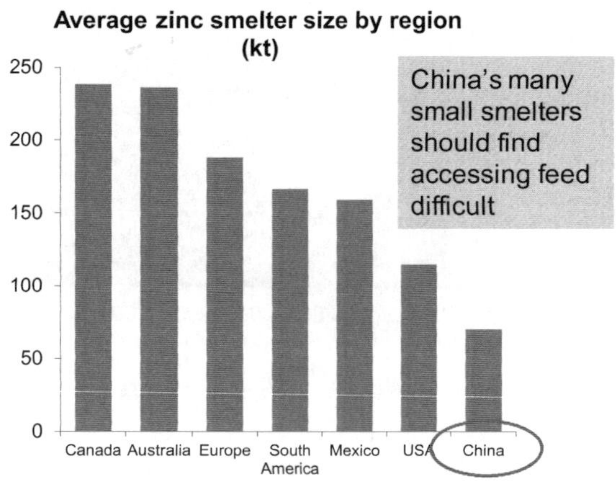

Source: SHFE, LME, CRU, Antaike, Metal Bulletin, Macquarie Research, February 2017

Page 8

Macquarie Research | Commodities

MACQUARIE

Entering Phase II: The metal deficit

Which indicators will tell us Chinese cuts are happening?

1. Ingot premia
2. SHFE: LME arb opening
3. Ingot imports

Source: Macquarie Research, February 2017

Page 9

Macquarie Research | Commodities

MACQUARIE

Entering Phase II: The metal deficit

Which indicators will tell us Chinese cuts are happening?

1. Ingot premia
2. SHFE: LME arb opening
3. Ingot imports

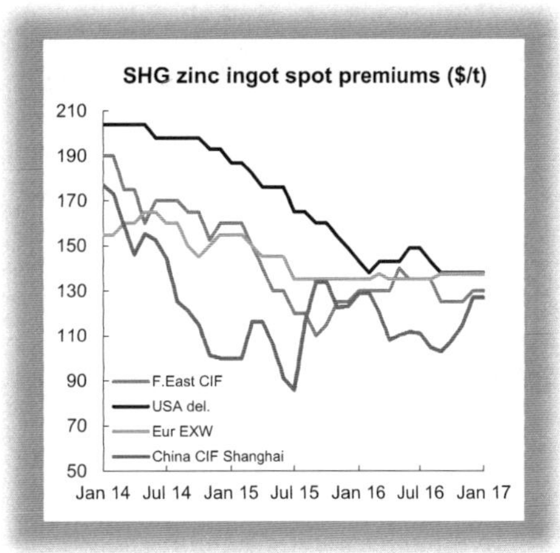

Source: CRU, Metal Bulletin, Macquarie Research, February 2017

Page 10

Entering Phase II: The metal deficit

Which indicators will tell us Chinese cuts are happening?

1. Ingot premia
2. SHFE: LME arb opening
3. Ingot imports

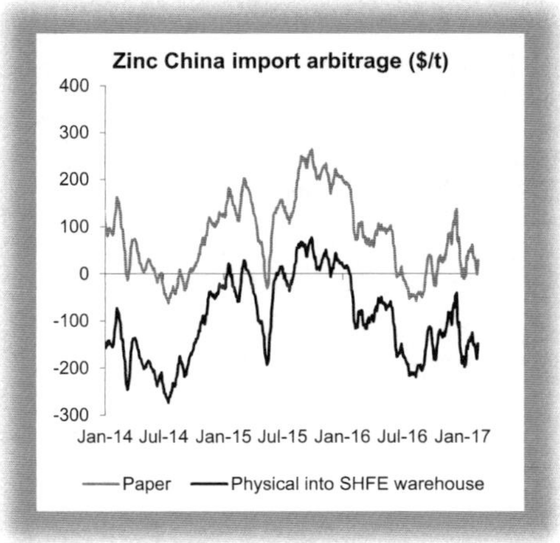

Source: SHFE, LME, CRU, Metal Bulletin, Macquarie Research, February 2017

Entering Phase II: The metal deficit

Which indicators will tell us Chinese cuts are happening?

1. Ingot premia
2. SHFE: LME arb opening
3. Ingot imports

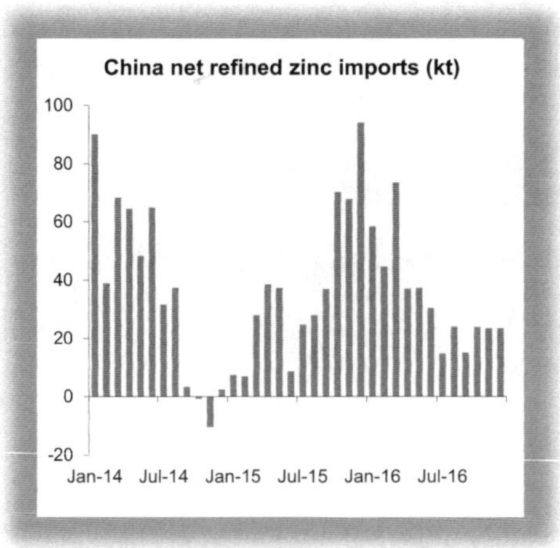

Source: China Customs, Macquarie Research, February 2017

Tightness thus seen supporting another leg up in prices

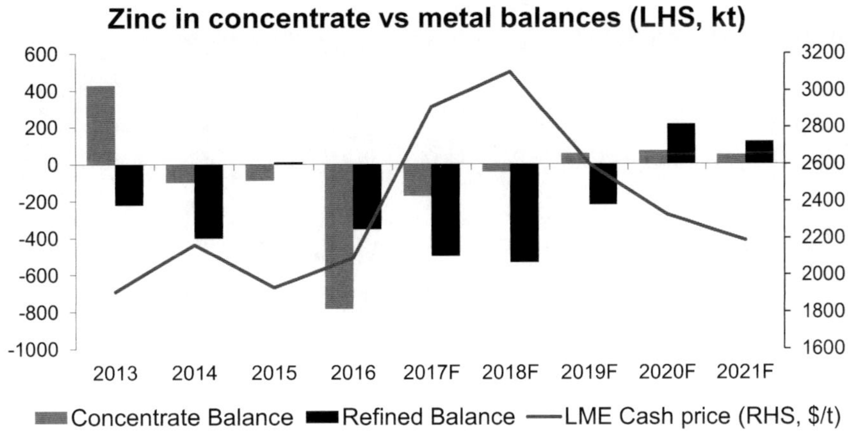

...but how do we return to balance later?

Source: Wood Mackenzie, CRU, Antaike, BGR Macquarie Research, February 2017

Page 13

The global industrial recovery is now maturing and becoming more broad-based

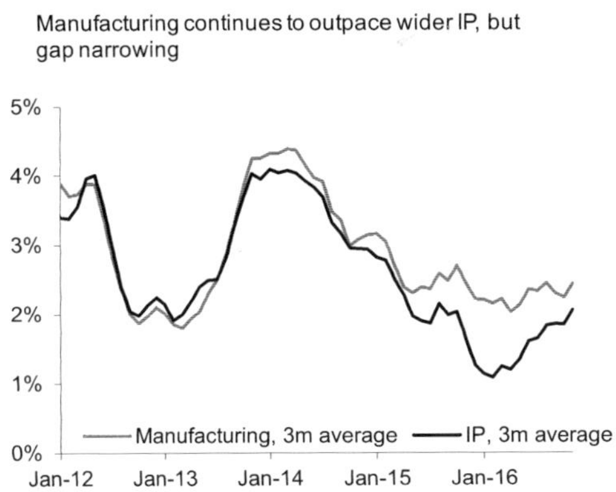

Source: National data, Markit, Macquarie Research, February 2017

Page 14

Macquarie Research | Commodities

MACQUARIE

Zinc demand currently stable, broad regional growth

Source: CRU, Macquarie Research, February 2017

Page 15

Macquarie Research | Commodities

MACQUARIE

Demand destruction…the key to the unwind

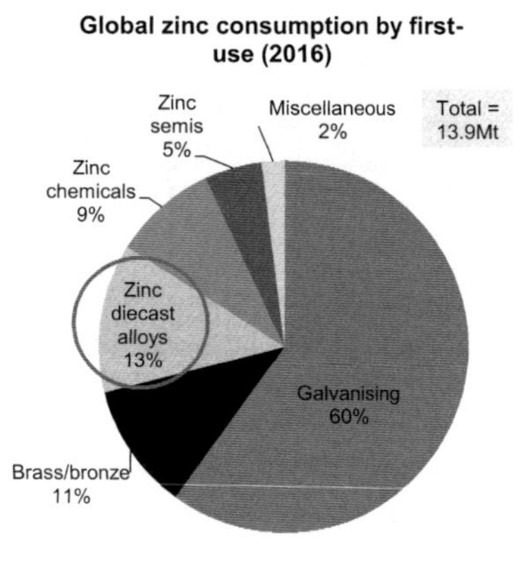

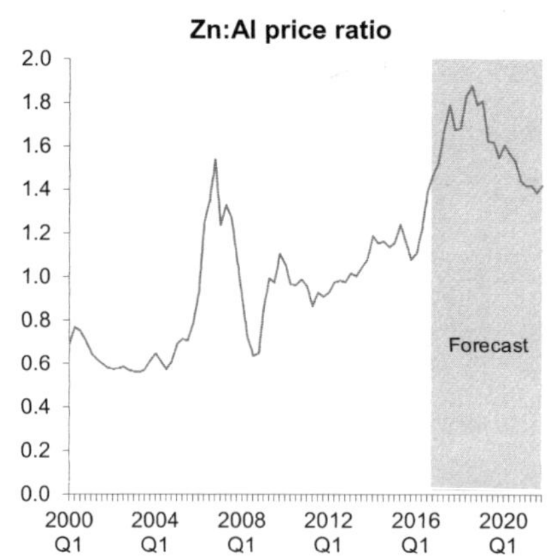

Source: LME, Wood Mackenzie, CRU, ILZSG, Macquarie Research, February 2017

Page 16

Macquarie Research | Commodities

Prices to step down from >$3,000/t as demand softens, mine projects kick in, and the market returns to balance

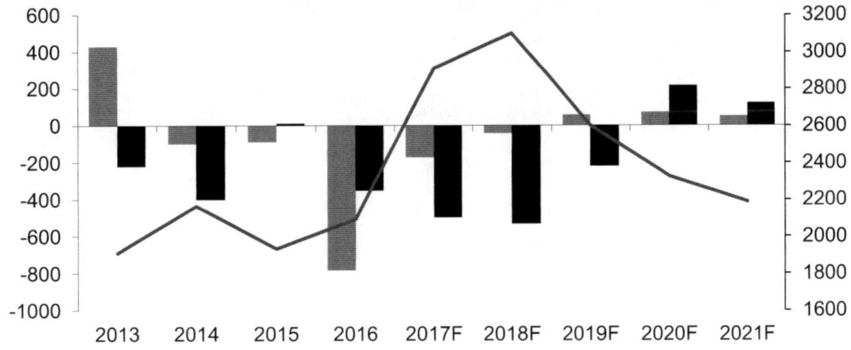

Source: Wood Mackenzie, CRU, Antaike, BGR Macquarie Research, February 2017

Page 17

Macquarie Research | Commodities

Thank you

Vivienne Lloyd

+44 20 3037 4530
vivienne.lloyd@macquarie.com

Macquarie Commodities Research
Macquarie Capital (Europe) Ltd
28 Ropemaker Street
London EC2Y 9HD, UK

February 2017

Page 18

Important disclosures:

Recommendation definitions

Macquarie - Australia/New Zealand

Outperform – return > 3% in excess of benchmark return
Neutral – return within 3% of benchmark return
Underperform – return > 3% below benchmark return

Benchmark return is determined by long term nominal GDP growth plus 12 month forward market dividend yield

Macquarie – Asia/Europe

Outperform – expected return >+10%
Neutral – expected return from -10% to +10%
Underperform – expected <-10%

Macquarie - South Africa

Outperform – return > 10% in excess of benchmark return
Neutral – return within 10% of benchmark return
Underperform – return > 10% below benchmark return

Macquarie - Canada

Outperform – return > 5% in excess of benchmark return
Neutral – return within 5% of benchmark return
Underperform – return > 5% below benchmark return

Macquarie - USA

Outperform – return > 5% in excess of benchmark return
Neutral – return within 5% of benchmark return
Underperform – return > 5% below benchmark return

Volatility index definition*
This is calculated from the volatility of historic price movements.

Very high–highest risk – Stock should be expected to move up or down 60-100% in a year – investors should be aware this stock is highly speculative.

High – stock should be expected to move up or down at least 40-60% in a year – investors should be aware this stock could be speculative.

Medium – stock should be expected to move up or down at least 30-40% in a year.

Low–medium – stock should be expected to move up or down at least 25-30% in a year.

Low – stock should be expected to move up or down at least 15-25% in a year.

*Applicable to Australian/NZ stocks only

Recommendation – 12 months

Note: Quant recommendations may differ from Fundamental Analyst recommendations

Financial definitions

All "Adjusted" data items have had the following adjustments made:

Added back: goodwill amortisation, provision for catastrophe reserves, IFRS derivatives & hedging, IFRS impairments & IFRS interest expense
Excluded: non recurring items, asset revals, property revals, appraisal value uplift, preference dividends & minority interests

EPS = adjusted net profit /efpowa*
ROA = adjusted ebit / average total assets
ROA Banks/Insurance = adjusted net profit /average total assets
ROE = adjusted net profit / average shareholders funds
Gross cashflow = adjusted net profit + depreciation
*equivalent fully paid ordinary weighted average number of shares

All Reported numbers for Australian/NZ listed stocks are modelled under IFRS (International Financial Reporting Standards).

Recommendation proportions – For quarter ending 31 December 2016

	AU/NZ	Asia	RSA	USA	CA	EUR	
Outperform	57.53%	50.72%	45.57%	42.28%	60.58%	52.79%	(for global coverage by Macquarie, 8.71% of stocks followed are investment banking clients)
Neutral	33.90%	33.97%	43.04%	50.11%	37.23%	35.62%	(for global coverage by Macquarie, 8.05% of stocks followed are investment banking clients)
Underperform	8.56%	15.30%	11.39%	7.61%	2.19%	11.59%	(for global coverage by Macquarie, 4.63% of stocks followed are investment banking clients)

Company-Specific Disclosures:
Important disclosure information regarding the subject companies covered in this report is available at www.macquarie.com/research/disclosures.

Analyst Certification:
We hereby certify that all of the views expressed in this report accurately reflect our personal views about the subject company or companies and its or their securities. We also certify that no part of our compensation was, is or will be, directly or indirectly, related to the specific recommendations or views expressed in this report. The Analysts responsible for preparing this report receive compensation from Macquarie that is based upon various factors including Macquarie Group Ltd total revenues, a portion of which are generated by Macquarie Group's Investment Banking activities.

General Disclaimers:
Macquarie Securities (Australia) Ltd; Macquarie Capital (Europe) Ltd; Macquarie Capital Markets Canada Ltd; Macquarie Capital Markets North America Ltd; Macquarie Capital (USA) Inc; Macquarie Capital Limited and Macquarie Capital Limited, Taiwan Securities Branch; Macquarie Capital Securities (Singapore) Pte Ltd; Macquarie Securities (NZ) Ltd; Macquarie Equities South Africa (Pty) Ltd.; Macquarie Capital Securities (India) Pvt Ltd; Macquarie Capital Securities (Malaysia) Sdn Bhd; Macquarie Securities Korea Limited and Macquarie Securities (Thailand) Ltd are not authorized deposit-taking institutions for the purposes of the Banking Act 1959 (Commonwealth of Australia), and their obligations do not represent deposits or other liabilities of Macquarie Bank Limited ABN 46 008 583 542 (MBL) or MGL. MBL does not guarantee or otherwise provide assurance in respect of the obligations of any of the above mentioned entities. MGL provides a guarantee to the Monetary Authority of Singapore in respect of the obligations and liabilities of Macquarie Capital Securities (Singapore) Pte Ltd for up to SGD 35 million. This research has been prepared for the general use of the wholesale clients of the Macquarie Group and must not be copied, either in whole or in part, or distributed to any other person. If you are not the intended recipient you must not use or disclose the information in this research in any way. If you received it in error, please tell us immediately by return e-mail and delete the document. We do not guarantee the integrity of any e-mails or attached files and are not responsible for any changes made to them by any other person. MGL has established and implemented a conflicts policy at group level (which may be revised and updated from time to time) (the "Conflicts Policy") pursuant to regulatory requirements (including the FCA Rules) which sets out how we must seek to identify and manage all material conflicts of interest. Nothing in this research shall be construed as a solicitation to buy or sell any security or product, or to engage in or refrain from engaging in any transaction. In preparing this research, we did not take into account your investment objectives, financial situation or particular needs. Macquarie salespeople, traders and other professionals may provide oral or written market commentary or trading strategies to our clients that reflect opinions which are contrary to the opinions expressed in this research. Macquarie Research produces a variety of research products including, but not limited to, fundamental analysis, macro-economic analysis, quantitative analysis, and trade ideas. Recommendations contained in one type of research product may differ from recommendations contained in other types of research, whether as a result of differing time horizons, methodologies, or otherwise. Before making an investment decision on the basis of this research, you need to consider, with or without the assistance of an adviser, whether the advice is appropriate in light of your particular investment needs, objectives and financial circumstances. There are risks involved in securities trading. The price of securities can and does fluctuate, and an individual security may even become valueless. International investors are reminded of the additional risks inherent in international investments, such as currency fluctuations and international stock market or economic conditions, which may adversely affect the value of the investment. This research is based on information obtained from sources believed to be reliable but we do not make any representation or warranty that it is accurate, complete or up to date. We accept no obligation to correct or update the information or opinions in it. Opinions expressed are subject to change without notice. No member of the Macquarie Group accepts any liability whatsoever for any direct, indirect, consequential or other loss arising from any use of this research and/or further communication in relation to this research. Clients should contact analysts at, and execute transactions through, a Macquarie Group entity in their home jurisdiction unless governing law permits otherwise. The date and timestamp for above share price and market cap is the closed price of the price date. #CLOSE is the final price at which the security is traded in the relevant exchange on the date indicated.

Country-Specific Disclaimers:
Australia: In Australia, research is issued and distributed by Macquarie Securities (Australia) Ltd (AFSL No. 238947), a participating organisation of the Australian Securities Exchange. **New Zealand**: In New Zealand, research is issued and distributed by Macquarie Securities (NZ) Ltd, a NZX Firm. **Canada**: In Canada, research is prepared, approved and distributed by Macquarie Capital Markets Canada Ltd, a participating organisation of the Toronto Stock Exchange, TSX Venture Exchange & Montréal Exchange. Macquarie Capital Markets North America Ltd., which is a registered broker-dealer and member of FINRA, accepts responsibility for the contents of reports issued by Macquarie Capital Markets Canada Ltd in the United States and sent to US persons. Any US person wishing to effect transactions in the securities described in the reports issued by Macquarie Capital Markets Canada Ltd should do so with Macquarie Capital Markets North America Ltd. The Research Distribution Policy of Macquarie Capital Markets Canada Ltd is to allow all clients that are entitled to have equal access to our research. **United Kingdom**: In the United Kingdom, research is issued and distributed by Macquarie Capital (Europe) Ltd, which is authorised and regulated by the Financial Conduct Authority (No. 193905). **Germany**: In Germany, this research is issued and/or distributed by Macquarie Capital (Europe) Limited, Niederlassung Deutschland, which is authorised and regulated by the UK Financial Conduct Authority (No. 193905). and in Germany by BaFin. **France**: In France, research is issued and distributed by Macquarie Capital (Europe) Ltd, which is authorised and regulated in the United Kingdom by the Financial Conduct Authority (No. 193905). **Hong Kong & Mainland China**: In Hong Kong, research is issued and distributed by Macquarie Capital Limited, which is licensed and regulated by the Securities and Futures Commission. In Mainland China, Macquarie Securities (Australia) Limited Shanghai Representative Office only engages in non-business operational activities excluding issuing and distributing research. Only non-A share research is distributed into Mainland China by Macquarie Capital Limited. **Japan**: In Japan, research is issued and distributed by Macquarie Capital Securities (Japan) Limited, a member of the Tokyo Stock Exchange, Inc. and Osaka Exchange, Inc. (Financial Instruments Firm, Kanto Financial Bureau (kin-sho) No. 231, a member of Japan Securities Dealers Association). **India**: In India, research is issued and distributed by Macquarie Capital Securities (India) Pvt. Ltd. (CIN: U65920MH1995PTC090696), 92, Level 9, 2 North Avenue, Maker Maxity, Bandra Kurla Complex, Bandra (East), Mumbai – 400 051, India, which is a SEBI registered Research Analyst having registration no. INH000000545. **Malaysia**: In Malaysia, research is issued and distributed by Macquarie Capital Securities (Malaysia) Sdn. Bhd. (Company registration number: 463469-W) which is a Participating Organisation of Bursa Malaysia Berhad and a holder of Capital Markets Services License issued by the Securities Commission. **Taiwan**: In Taiwan, research is issued and distributed by Macquarie Capital Limited, Taiwan Securities Branch, which is licensed and regulated by the Financial Supervisory Commission. No portion of the report may be reproduced or quoted by the press or any other person without authorisation from Macquarie. Nothing in this research shall be construed as a solicitation to buy or sell any security or product. The recipient of this report shall not engage in any activities which may give rise to potential conflicts of interest to the report. Research Associate(s) in this report who are registered as Clerks only assist in the preparation of research and are not engaged in writing the research. **Thailand**: In Thailand, research is produced, issued and distributed by Macquarie Securities (Thailand) Ltd. Macquarie Securities (Thailand) Ltd. is a licensed securities company that is authorized by the Ministry of Finance, regulated by the Securities and Exchange Commission of Thailand and is an exchange member of the Stock Exchange of Thailand. The Thai Institute of Directors Association has disclosed the Corporate Governance Report of Thai Listed Companies made pursuant to the policy of the Securities and Exchange Commission of Thailand. Macquarie Securities (Thailand) Ltd does not endorse the result of the Corporate Governance Report of Thai Listed Companies but this Report can be accessed at: http://www.thai-iod.com/en/publications.asp?type=4. **South Korea**: In South Korea, unless otherwise stated, research is prepared, issued and distributed by Macquarie Securities Korea Limited, which is regulated by the Financial Supervisory Services. Information on analysts in MSKL is disclosed at: http://dis.kofia.or.kr/websquare/index.jsp?w2xPath=/wq/fundMgr/DISFundMgrAnalystStut.xml&divisionId=MDIS03002001000000&serviceId=SDIS03002001000. **South Africa**: In South Africa, research is issued and distributed by Macquarie Equities South Africa (Pty) Ltd., a member of the JSE Limited. **Singapore**: In Singapore, research is issued and distributed by Macquarie Capital Securities (Singapore) Pte Ltd (Company Registration Number: 198702912C), a Capital Markets Services license holder under the Securities and Futures Act to deal in securities and provide custodial services in Singapore. Pursuant to the Financial Advisers (Amendment) Regulations 2005, Macquarie Capital Securities (Singapore) Pte Ltd is exempt from complying with sections 25, 27 and 36 of the Financial Advisers Act. All Singapore-based recipients of research produced by Macquarie Capital (Europe) Limited, Macquarie Capital Markets Canada Ltd, Macquarie Equities South Africa (Pty) Ltd.and Macquarie Capital (USA) Inc. represent and warrant that they are institutional investors as defined in the Securities and Futures Act. **United States**: In the United States, research is issued and distributed by Macquarie Capital (USA) Inc., which is a registered broker-dealer and member of FINRA. Macquarie Capital (USA) Inc, accepts responsibility for the content of each research report prepared by one of its non-US affiliates when the research report is distributed in the United States by Macquarie Capital (USA) Inc. Macquarie Capital (USA) Inc.'s affiliate analysts are not registered as research analysts with FINRA, may not be associated persons of Macquarie Capital (USA) Inc., and therefore may not be subject to FINRA rule restrictions on communications with a subject company, public appearances, and trading securities held by a research analyst account. Information regarding futures is provided for reference purposes only and is not a solicitation for purchases or sales of futures. Any persons receiving this report directly from Macquarie Capital (USA) Inc. and wishing to effect a transaction in any security described herein should do so with Macquarie Capital (USA) Inc. Important disclosure information regarding the subject companies covered in this report is available at www.macquarie.com/research/disclosures, or contact your registered representative at 1-888-MAC-STOCK, or write to the Supervisory Analysts, Research Department, Macquarie Securities, 125 W.55th Street, New York, NY 10019.

Contact Macquarie Capital (USA) Inc.

SALES

US

Khristina McLaughlin	(212) 231 8012
Michael Burguieres	(212) 231 2506
Michael Cline	(212) 231 2524
Abigail Lang	(212) 231 0937
Alex Martin	(212) 231 0376
Khalid McCaskill	(212) 231 2505
Robert McKeon	(212) 231 2573
Rob Moderelli	(212) 231 2495
Heidi Muccifori	(212) 231 2466
Tad Nacheff	(212) 231 2615
Mark Orsi	(415) 762 5002
Shaival Patel	(212) 231 8002
Jason Philosophos	(212) 231 2567
Troy Pilalas	(415) 762 5025
Jason Rubin	(212) 231 8011
Denver Smith	(617) 598 2507
Lisa Sowell	(212) 231 1787

Australia/NZ

Leighton Patrick	(212) 231 2552
Michael Cowcher	(212) 231 2478
Michael McNair	(212) 231 2571

Asia

Eric Roles	(212) 231 2559
Shunsuke Aonuma	(617) 598 2535
Paul Colaco	(415) 762 5003
Jeff Evans	(617) 598 2508
Mark Lawrence	(212) 231 2516
Todd Narter	(415) 762 5005
Brad Neill	(212) 231 0973
Jeff Seo	(212) 231 2422
Jean Zhang	(212) 231 6397

Europe

Christian Schmuck	(212) 231 8007
Chris Carr	(212) 231 6398
Jan Halaska	(617) 598 2503
Doug Stone	(212) 231 2606

South Africa

Jesse Ushewokunze	(212) 231 2504

SALES TRADING

US

J.T. Cacciabaudo	(212) 231 6381
Chris Baker	(212) 231 8064
Steven Belge	(617) 598 2515
Greg Calvino	(212) 231 0902
Frank Casey	(617) 598 2509
Mike Gray	(212) 231 2555
Brad Hoenig	(212) 231 8089
Pat Jackson	(212) 231 8043
Carlos Jimenez	(212) 231 0883
Allen Jordan	(212) 231 0947
Marty Livingston	(212) 231 2523
Matt McCloskey	(212) 231 8024
Taylor Ortiz	(415) 762 5059
Rob Salmon	(415) 762 5009
Patrick Scanlan	(312) 660 9025
Tarik Turner	(212) 231 6365

International

Chris Reale	(212) 231 2555
Valerie Krah	(212) 231 2555

TRADING

Brian Ross	(212) 231 2534
Ed Bartleson	(212) 231 2555
Pete McQuaid	(212) 231 2459
Michael Nevitt	(212) 231 2408
Chris Zarins	(212) 231 0882

EXECUTION SERVICES (PT)

Matt McNamara	(212) 231 0900
Ben Salup	(212) 231 0900
Jessica Pepin	(212) 231 0900

EXECUTION SERVICES (EE)

Stuart McWilliam	(212) 231 0925
Matt Hor	(212) 231 0300
Shaun McDonnell	(212) 231 0300
Marc Rosa	(212) 231 0300

CORPORATE ACCESS

Margaret Shergalis	(212) 231 1348
Katie Gray	(212) 231 1318
Karen Bader	(212) 231 6364
Alexa Sapper	(212) 231 1768

Email Addresses
firstname.surname@macquarie.com
eg robert.ansell@macquarie.com

Macquarie Capital (USA) Inc. 125 W. 55th Street, Level 22 New York, NY 10019	Macquarie Capital (USA) Inc. 2 Embarcadero Center, Suite 200 San Francisco, CA 94111	Macquarie Capital (USA) Inc. 225 Franklin Street, 17th Floor Boston, MA 02110	Macquarie Capital (USA) Inc. 1 North Wacker Drive, Level 29 Chicago, IL 60606	Macquarie Capital (USA) Inc 1225 17th Street, Suite 1600 Denver, CO, 80202	24 Hour Toll-Free 1-888 MAC STOCK 1-888 622 7862
(212) 231 1000	(415) 762 5000	(617) 598 2500	(312) 660 9215	(303) 952 2800	

The role of physical trading firms in the zinc market

IZA Conference

Palm Springs – 28 February 2017

Concord Resources Limited

Duncan Hobbs
+44 20 3879 3966
duncan.hobbs@concordltd.com

CONCORD

Disclaimer

Concord Resources Limited
Who we are

- Concord is a global resources merchant focused on the supply chain and risk management of non-ferrous metals and associated raw materials, led by an experienced and entrepreneurial management team
- Concord is a global company, headquartered in London and with fully-staffed offices in New York and Hong Kong. Our experienced team work with resource producers and consumers on spot transactions out to multi-year, structured agreements
- Concord's capabilities include global sourcing, marketing and management of all required logistics and finance to optimise our clients' supply chains from long-term partnership marketing arrangements to single container deliveries
- Concord brings considerable cross-commodity, structuring and asset optimisation experience of its founding team to work for the benefit of its customers and partners
- Concord is an independent company, which combines its founding management's experience with the commodity and financial industry expertise of its shareholders, including:
  - MKS Switzerland SA
  - Pala Investments Limited
  - Principals from Ospraie Management
  - Pinnacle Asset Management

CONCORD 2

The role of physical trading firms in the zinc market

- Who?
- Why?
- What?
- How?
- Summary
- Concord

CONCORD 3

The role of physical trading firms in the zinc market

- Who?
- Why?
- What?
- How?
- Summary
- Concord

CONCORD 4

Physical trading firms active in non-ferrous metals markets are highly diverse

- **Size**
  From small-scale sole traders to multi-national multi-billion dollar firms
- **Product range**
  From focusing on a single market or small group of markets, such as base metals, to wide-ranging activity across agricultural, energy, bulks and metals markets
- **Geographic range**
  From regional distributors to global operators
- **Ownership**
  Some are public companies,, some are private and some have hybrid structures
- **Operations**
  From "asset light" to "asset heavy" with own mining and smelting operations
- **Structure**
  Some are independent; some are owned by producers, consumers or banks

Source: Concord Resources Limited, February 2017

CONCORD 5

Some of the physical trading firms active in the world markets for zinc concentrates and refined metal

	Company	Zinc concentrates	Zinc metal
1.	Castleton Commodities International (CCI)		✓
2.	Cliveden Trading AG	✓	
3.	Concord Resources Limited	✓	✓
4.	Freepoint Commodities	✓	✓
5.	Gerald Group		✓
6.	Glencore	✓	✓
7.	Louis Dreyfus Company	✓	✓
8.	MRI Group	✓	
9.	Nizi International		✓
10.	Noble Group		✓
11.	Ocean Partners	✓	
12.	Penfold World Trade	✓	
13.	Trafigura	✓	✓
14.	Transamine	✓	
15.	Traxys	✓	✓

Note: 1. Firms ex-China. This list, organised alphabetically, is intended to be indicative and not exhaustive.
Source: Websites, Concord Resources Limited, February 2017

CONCORD 6

The role of physical trading firms in the zinc market

- Who?
- Why?
- What?
- How?
- Summary
- Concord

CONCORD 7

Physical trading firms' activities are in part a function of the realities of the physical world

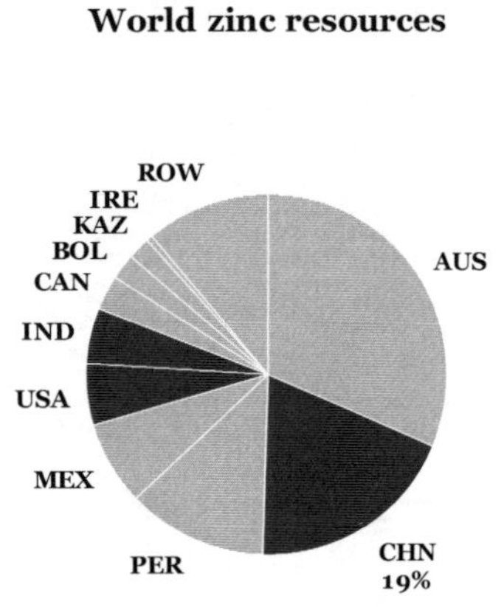

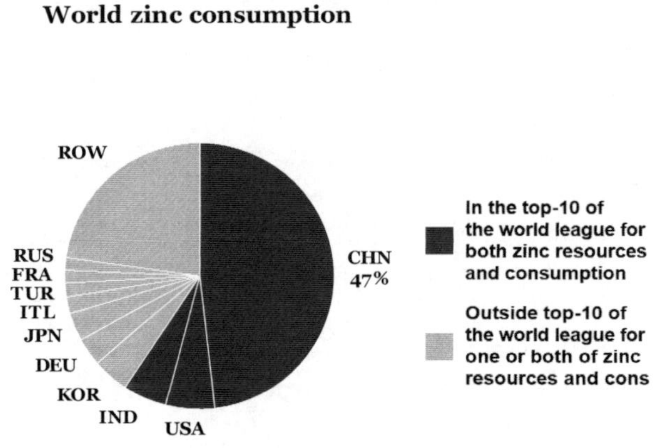

- Where resources occur in nature is different from where resources are wanted in the world

Note: 2015 data for resources. 2016 data for consumption.
Source: ILZSG, USGS, Concord Resources Limited, February 2017

CONCORD 8

China has had to increase substantially its imports of zinc concentrates to meet its rising demand for refined metal output

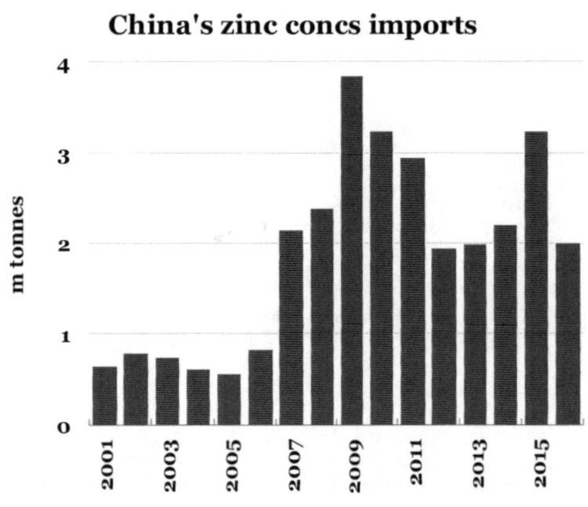

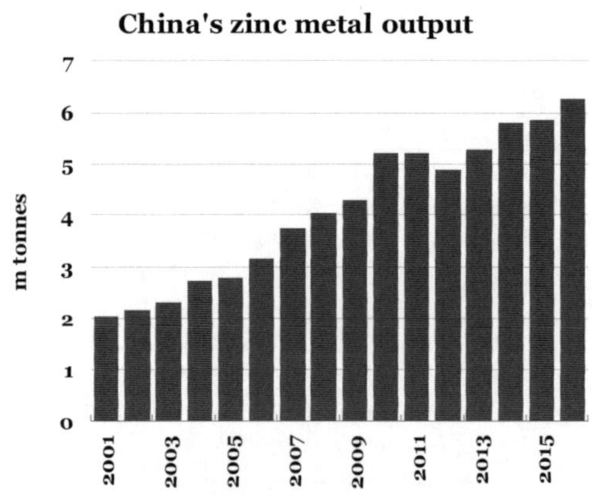

- China's imports of zinc concentrates increased almost fivefold from 2001-2015
- In 2016 fall in imports due to reduced supplies, *not* declining demand

Note: All data are gross weight, not metal contained.
Source: China Customs, Concord Resources Limited, February 2017

CONCORD 9

Physical trading firms are responding to the inherent structure of the industries they serve

- Zinc mining and smelting are capital intensive industries, operating margins are often lean and the time horizons over which assets can be expected to operate extend well beyond anything that can be reasonably predicted
- From an operational perspective, process plants often perform best when running close to full design capacity; lower operating rates can incur significant unit cost penalties, with obvious implications for margins and earnings
- But the business cycle is highly variable, which can and does lead to mismatches between the level of output appropriate to market demand vs. optimal operational performance of mines / smelters
- Mining is an uncertain activity since conditions can vary widely and unexpectedly, while process plants can breakdown, as a result of which producers will usually commit to sell less than planned production / full output on contract

- **Mismatches between planned / actual zinc output and market demand are common**

Source: Concord Resources Limited, February 2017

For example, one of the world's largest zinc mining and smelting firms has seen sharp swings in output over the last year

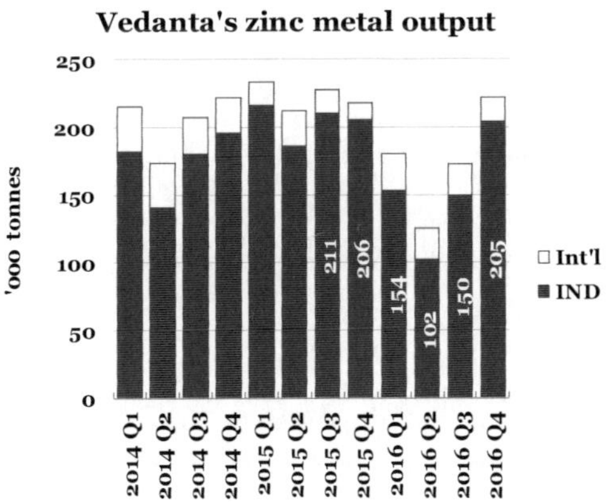

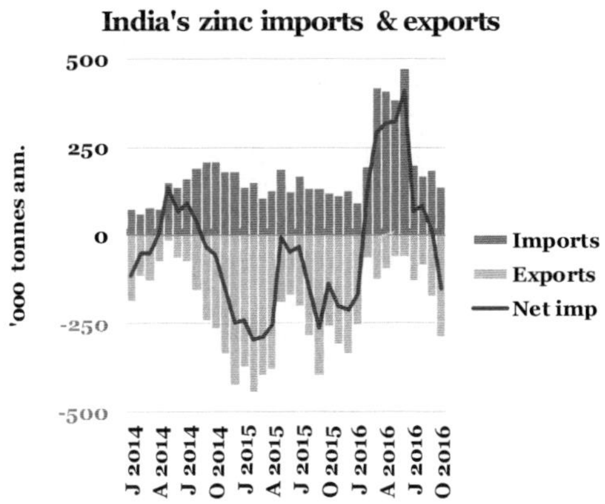

- Vedanta's Indian zinc metal output down -39% in H1 2016 vs. H2 2015; imports up +148%
- Source of increased imports strongly suggests market response facilitated by physical traders

Note: All data basis calendar years. Import data basis HSC7901.
Source: Indian Customs, Vedanta Resources, Concord Resources Limited, February 2017

The role of physical trading firms in the zinc market

- Who?
- Why?
- What?
- How?
- Summary
- Concord

CONCORD

Physical trading firms add value by making product modifications through difference(s) (arbitrage) in time, space and / or form

- **Time** – for example, storage of surplus supplies
- **Space** – for example, transfer from regions of surplus to deficit
- **Form** – for example, blending of concentrates; aggregation / disaggregation of cargoes
- There may be constraints on what and how product modifications might be made:
  - Physical – for example, related to infrastructure
  - Regulatory – for example, import / export restrictions such as for concentrates into China
  - Constraints can be time related – for example, time to put in place new cargo handling facilities
  - Typically the tighter the constraints, the higher the cost of making product modifications
- **Physical traders seek to optimise allocation and resolve constraints in markets**

Source: Concord Resources Limited, February 2017

CONCORD

Modifications in time

- This is mainly about the storage of material over time
- Modifications in time help zinc producers and consumers manage natural variance in activity arising from seasonality and the business cycle, and resulting differences in the timing of production and consumption
  - Storing metal and concentrates in times of surplus; releasing supplies in times of deficit
  - *"Base metals can always find a buyer [in] weak times – sales to traders and via the LME"* Boliden CEO, 2010
- Modifications in time also include "storage" in transit, moving materials such as concentrates from where resources occur to processing points, and refined metal from processing points to population centres, a necessary and integral part of the zinc supply chain
- Nature of materials has obvious implications for cost and duration of storage opportunities
  - Refined zinc – high-value, low-volume, non-perishable
  - Concentrates may require more careful management in storage

Source: Concord Resources Limited, February 2017

CONCORD 14

Modifications in space

- TCs for zinc concentrates and premiums for metal vary by region, according to the local supply-demand balance and market structure (net import vs. export)
- *Note, however, data providers do not always report these on the same terms*
- Metal premiums are a function of:
  1. Delivery location (and date)
  2. Product form and specification
  3. Producer brand
  4. Parcel size
  5. Payment terms
  6. And Incoterms!
- Similar for concentrates

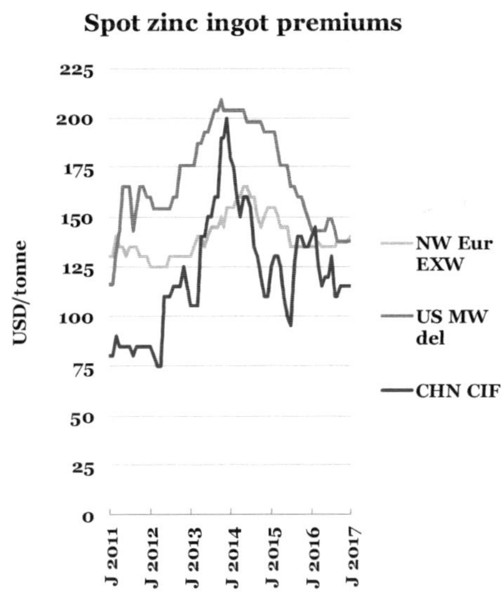

Source: CRU, MB, Concord Resources Limited, February 2017

CONCORD 15

Modifications in form

- Examples of modifications in form:

1. Aggregation of cargoes for deep sea shipment
   - Typically min. concs cargo ±5,000 wmt
   - ~3% of zinc mine output ex-China comes from mines producing <15,000 wmt pa, for which 5,000 wmt = 4 months' output
2. Disaggregation of cargoes for local distribution
3. Blending of zinc concentrates to manage impurities and physical properties
4. Remelting and casting of zinc metal in different forms

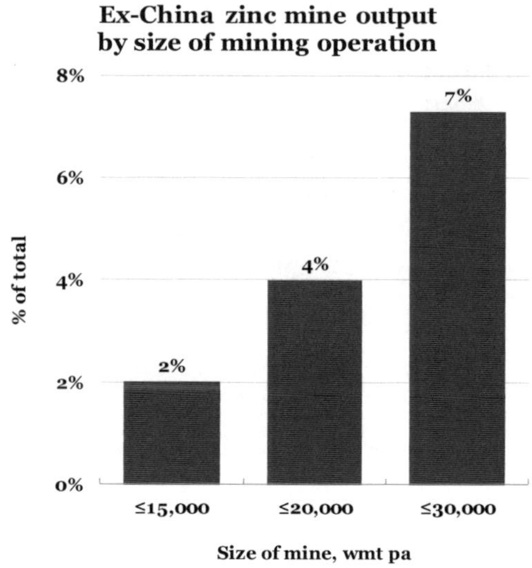

Note: 2016 data.
Source: Concord Resources Limited, February 2017

CONCORD

Physical trading firms also manage risk and complexity, provide flexibility and contribute competition

- Scope for product modifications provide opportunities for flexibility but also generate complexity and risks
- Physical traders may make multiple modifications to product(s) in time, space and form
- Zinc concentrates are complex products; >100 distinct qualities are reported to be in circulation in the world market
- There are myriad risks in the zinc market:
  - Flat price / basis / spread risk
  - Operational and performance risk
  - Credit / counterparty / currency risk
  - Legal and political risk
- These risks can be managed by various means including hedging, insurance, diversification, market knowledge and information
- Physical traders bidding for / offering products in the market contributes competition

Source: Concord Resources Limited, February 2017

CONCORD

The role of physical trading firms in the zinc market

- Who?
- Why?
- What?
- How?
- Summary
- Concord

Arbitrage

- Arbitrage is the avenue by which physical trading firms execute their business and add value for their customers
- Arbitrage opportunities arise where the value of a modification(s) in a product, as shown by the difference in prices before and after such modification has been made, is more than the cost of making the modification
- For example:
  - The TC for a parcel of blended zinc concentrates is lower than the weighted average of the separate materials combined in the blend
  - The premium for a parcel of metal delivered at one time and place is higher than premium prevailing at another time and place

Source: Concord Resources Limited, February 2017

Physical trading firms work with a number of variables between buying and selling terms in the zinc market

Concentrates
- TCs
- Metal payable
- Price participation (escalators / de-escalators)
- RCs and payables for precious metals
- Penalty charges for tramp elements

Refined metal
- Premiums
- Brands (listed vs. non-listed)
- Exchange prices (LME vs. SHFE)

Concentrates and metal
- Quotation periods (QPs) for pricing (may be v.valuable, esp. ± month of delivery)
- Ocean / inland freight terms / costs
- Payment terms (credit)

- **Outright level of TCs or premiums matters less to traders than miners or smelters**

Source: Concord Resources Limited, February 2017

CONCORD 20

Physical trading is typically a large revenue, lean margin business

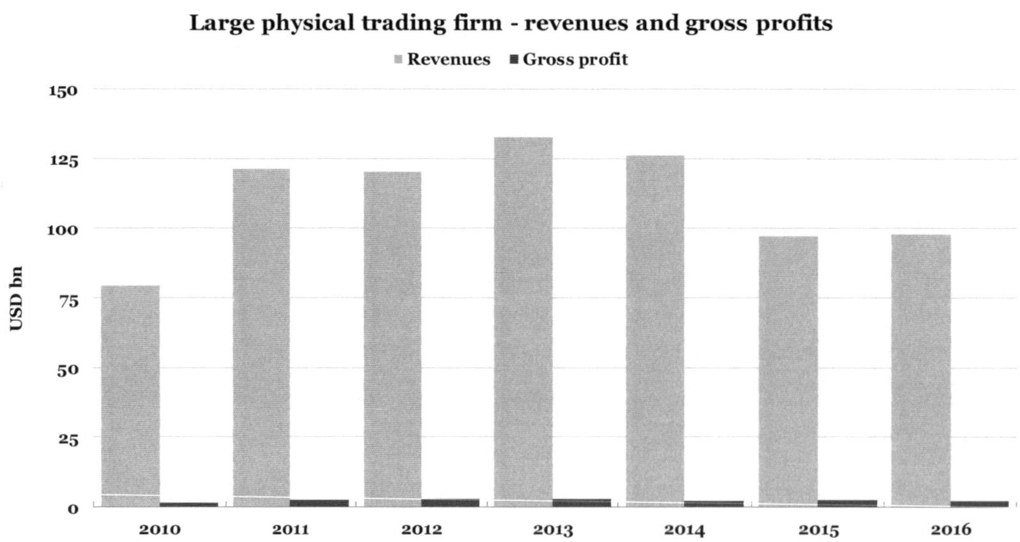

Note: For financial years ending September.
Source: Trafigura Group Pte Ltd, Concord Resources Limited, February 2017

CONCORD 21

Soft skills and funding

Soft skills

- **Information**: physical trading in zinc markets is an information intensive business
- **Knowledge**: industry knowledge and experience are invaluable in interpreting information
- **People**: this is a "people" business – networks are central to securing sourcing and sales
- **Attention to detail**: critical to effective execution (maybe difference between profit / loss)

Funding

- Physical trading firms draw on a wide range of sources to fund their business; secured and unsecured debt, equity, etc
- Funding arrangements will be a function of the modifications traders undertaken, notably assets employed and time take to make modifications
  - Short-term assets such as stocks will typically be funded with short-term debt
  - Long-term assets (warehouses, processing plants) typically funded with longer term debt

Source: Concord Resources Limited, February 2017

CONCORD 22

The role of physical trading firms in the zinc market

- Who?
- Why?
- What?
- How?
- Summary
- Concord

CONCORD 23

The role of physical trading firms in the zinc market

- Physical trading firms add value by making product modifications through difference(s) in time, space and / or form, within bounds of physical, regulatory and other constraints
- Physical trading firms manage risk and complexity, provide flexibility and contribute competition, bringing supplies to market that might otherwise be absent and helping to resolve mismatches between optimal mining and smelting operations and variable market conditions
- Arbitrage is the way by which physical trading firms execute their business and add value for their customers, where the value of product modification(s) as shown in price differences before and after modification is more than the cost of the modification, working with many variables between buying and selling terms
- Physical trading firms do not typically focus on flat price speculation and outright price levels matter less for traders than producers and consumers
- **Physical traders are an integral and invaluable part of zinc supply chains, without which producer / consumer working capital commitments would likely be higher, with important implications for earnings / end market sales prices**

Source: Concord Resources Limited, February 2017

CONCORD 24

The role of physical trading firms in the zinc market

- Who?
- Why?
- What?
- How?
- Summary
- Concord

CONCORD 25

Concord Resources Limited
Our markets: every step of the supply chain

	Aluminum	Copper	Zinc, lead, nickel	Alumina, bauxite	Precious metals
Metals				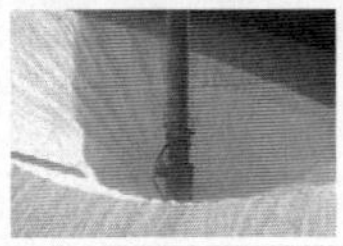	
Products	• P1020 • Billets • Alloys • Rod	• Cathode • Copper rod • Concentrates	• SHG & CGG zinc • 397 & 49s lead • Primary nickel • Concentrates	• SGA • Raw bauxite	• Gold • Silver • PGMs
Origination	• Consignments • Prepays/financing • Off-take/marketing • Tolling	• Consignments • Prepays/financing • Off-take/marketing • Tolling	• Consignments • Prepays/financing • Off-take/marketing • Tolling	• Swaps • Prepays/financing • Off-take/marketing • Tolling	• Off-take/marketing • By-products from concentrates
Storage / logistics	• LME system • Warehousing • Shipping	• LME system • Warehousing • Shipping • Blending	• LME system • Warehousing • Shipping • Blending	• Warehousing • Shipping • Blending	• Vaulting
Consumers	• Aerospace • Automotive • Construction • Packaging	• Wire and cable • Tubing • Brass • Smelters	• Galvanizers • Battery makers • Stainless & alloy steelmakers	• Refineries • Smelters	• Banks • Smelters

CONCORD

26

Concord Resources Limited
Contact our marketing teams

New York

437 MADISON AVENUE
28TH FLOOR
NEW YORK, NY 10022
USA

Scott Evans
Senior Trader Americas
Tel: +1 212 822 9793
scott.evans@concordltd.com

London

3 ST JAMES'S SQUARE
LEVEL 8
LONDON, SW1Y 4JU
ENGLAND

David Freeland
Senior Trader EMEA
Tel: +44 20 3879 3995
david.freeland@concordltd.com

Hong Kong

41 CONNAUGHT ROAD CENTRAL
LEVEL 16
NEXXUS BUILDING
HONG KONG

Felix Cauro
Senior Trader Asia
Tel: +852 3757 9915
felix.cauro@concordltd.com

CONCORD

27

IHS Markit

Presentation

2017 International Zinc Conference and Zinc Oxide Industry Conference

Zinc Chemical Market Trends

March 2017 ihsmarkit.com

Sean Davis, Principal Analyst, +1 408 343 4808, Sean.Davis@ihsmarkit.com

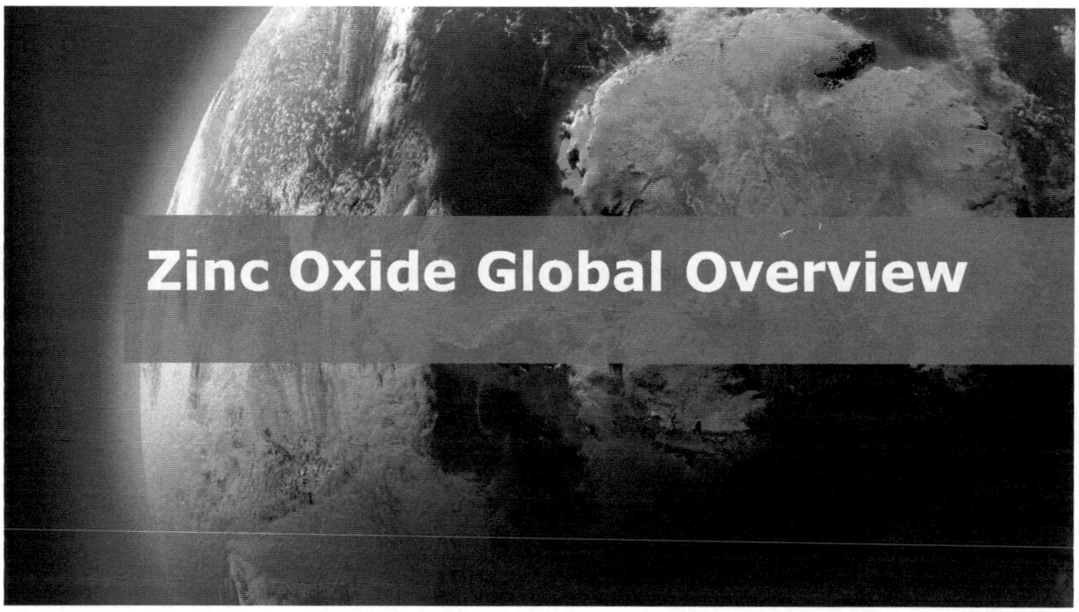

Zinc Oxide oversupplied globally

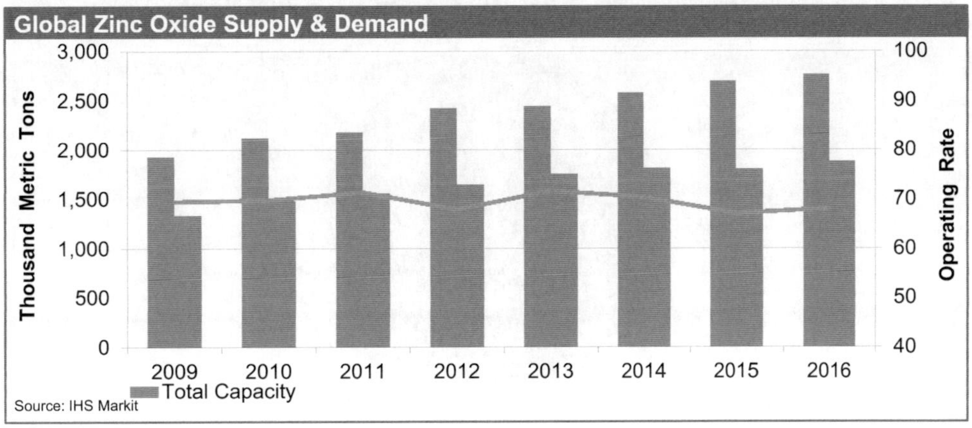

Source: IHS Markit

Northeast Asia consumes half

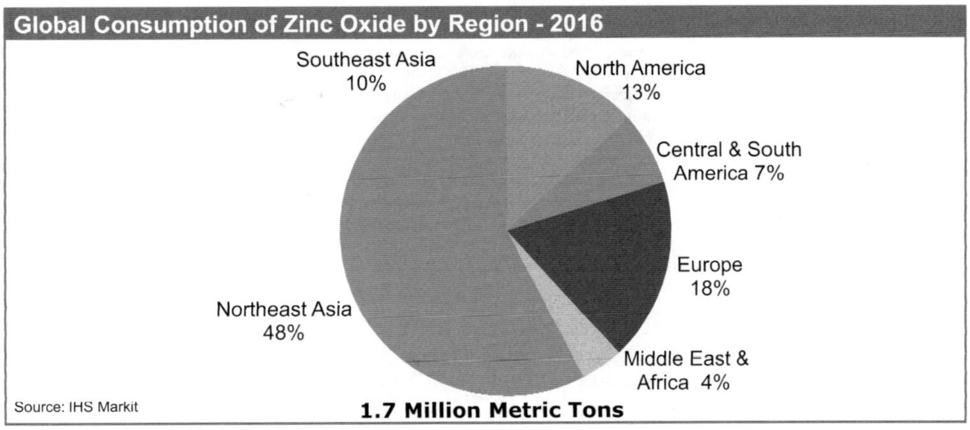

Source: IHS Markit

Rubber, Glass & Ceramics are Primary Uses

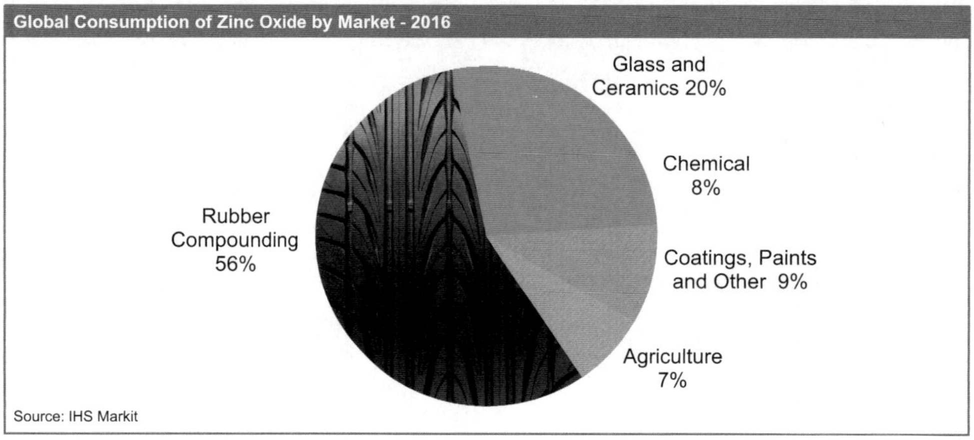

N-Rubber demand in Tires increasing

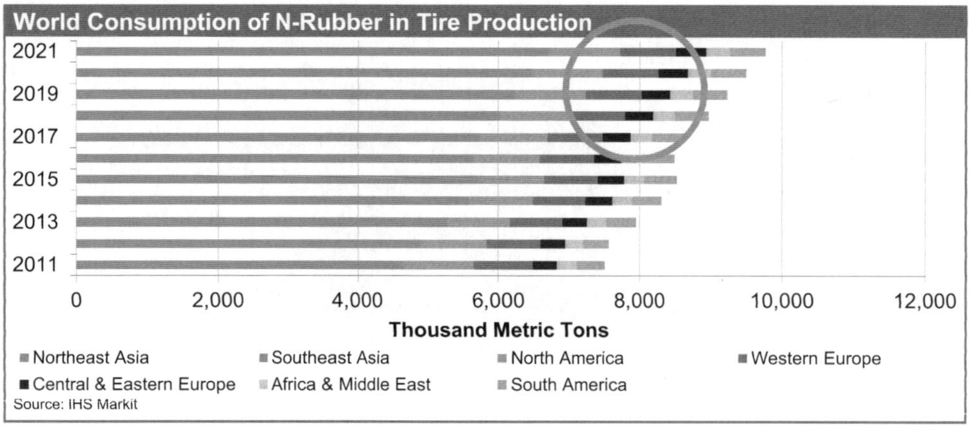

Tire capacity continues to shift

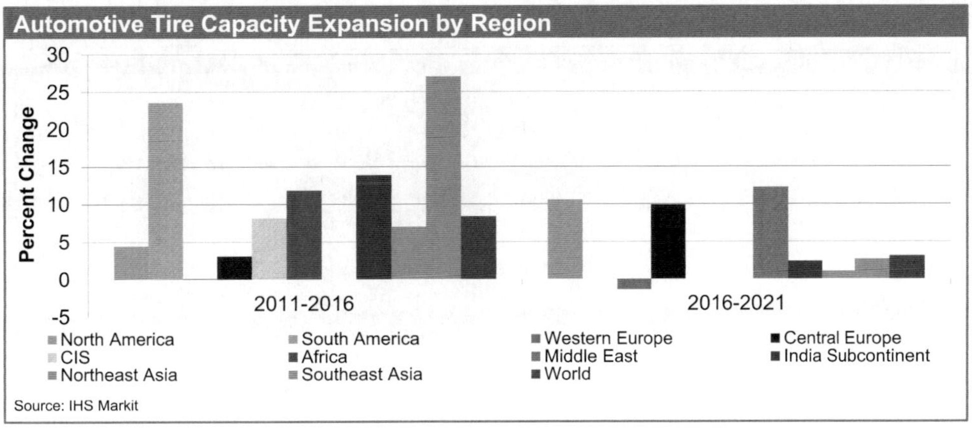

 7

N-Rubber demand in General Goods

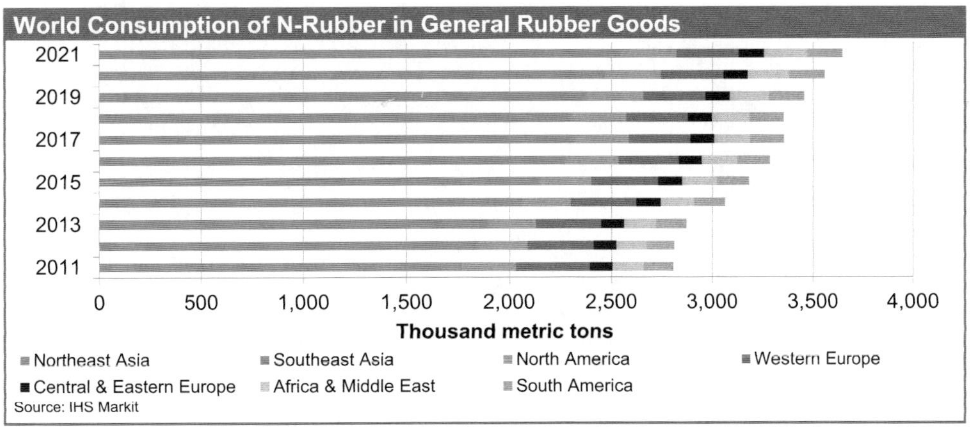

 8

As with rubber, Asia dominates ceramic and glass markets

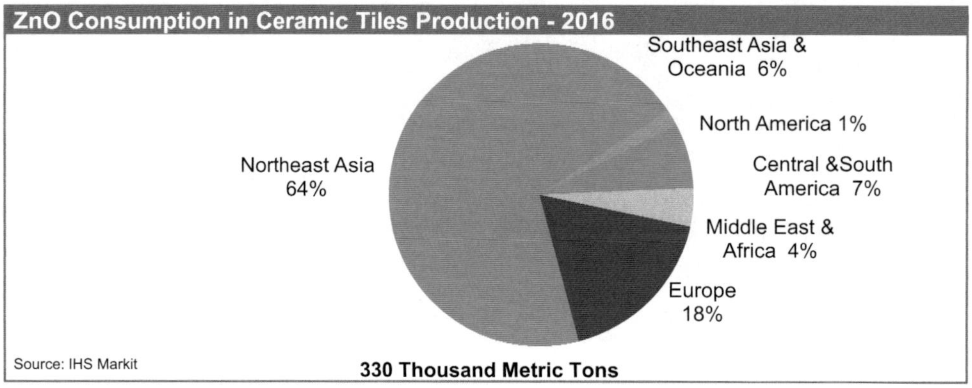

Where is ceramic tile produced?

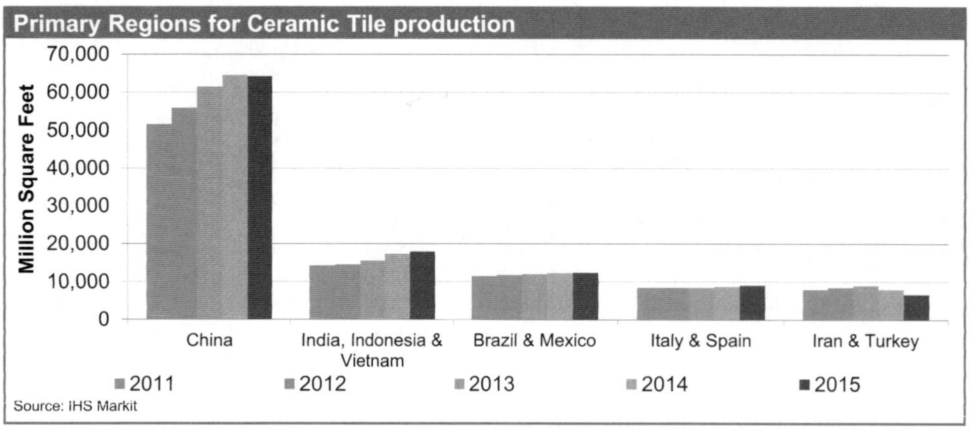

Consumption in Paints & Coatings are small

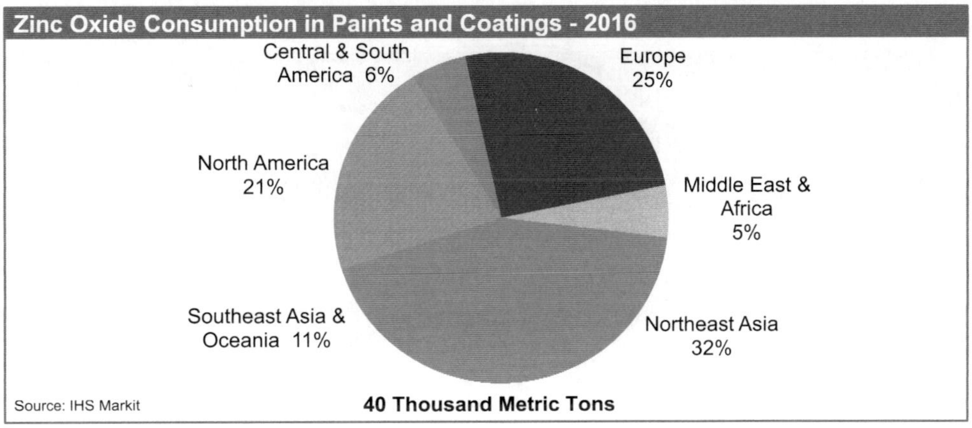

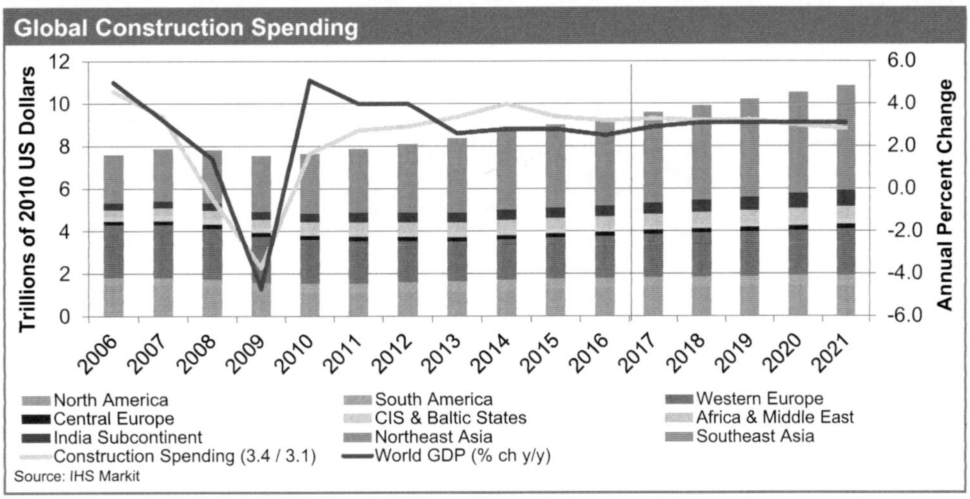

Zinc Oxide Consumption in Chemical Production

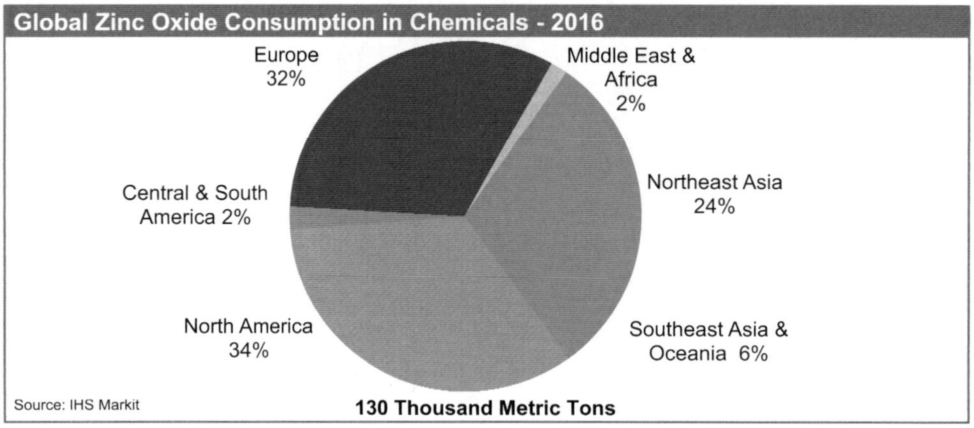

A starting material for zinc chemicals

Zinc oxide

Zinc borate

Zinc stearate

ZDBC / ZMBT

Zinc acetate

Zinc bromide

Zinc octoate

Zinc chloride

Zinc Naphthenate

Zinc phosphates (ZDDP)

Major micronutrient source in agriculture

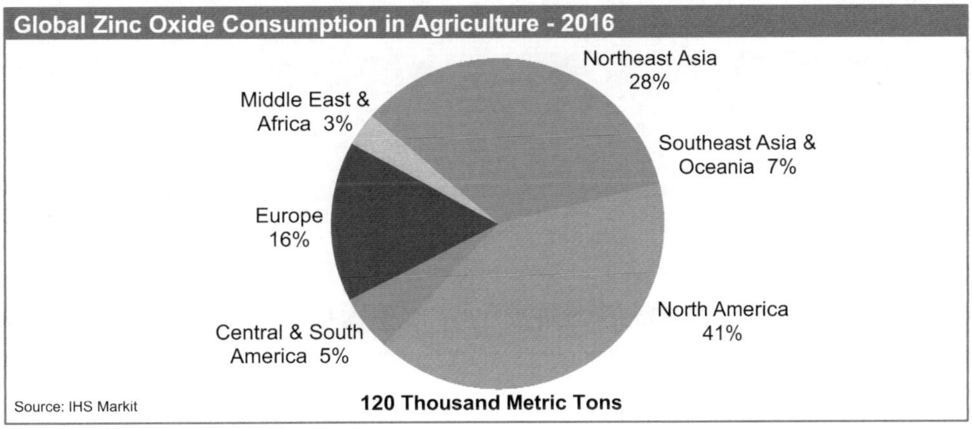

Demand in electronics remain steady

Closing Remarks

Trade of zinc chemicals

Tariffs on Intermediates and Finished Goods

Impact of potential policy changes in North America

The Technology of the New Mainstream

Zinc Oxide in Specialty Nutrition

Paul Reising
Compass Minerals
Wolf Trax Product Manager

Compass Minerals

Plant Nutrition

PROTASSIUM+ wolftrax Nu-Trax P+

What is Specialty Plant Nutrition?

Our definition:

- Innovative nutrients solutions
- Add value
- Provide extra benefit:
  - convenience
  - enhanced fertilizer performance
  - efficient use of time and resources

Our Focus

Creating better nutrient solutions for farmers around the world

Future ?

Will the future Specialty Nutrition products blur the line between "biological" and conventional fertilizer?

Compass Minerals | Plant Nutrition

Significant Opportunities in Mainstream U.S. Ag for Specialty Plant Nutrition

Favorable Market Trends

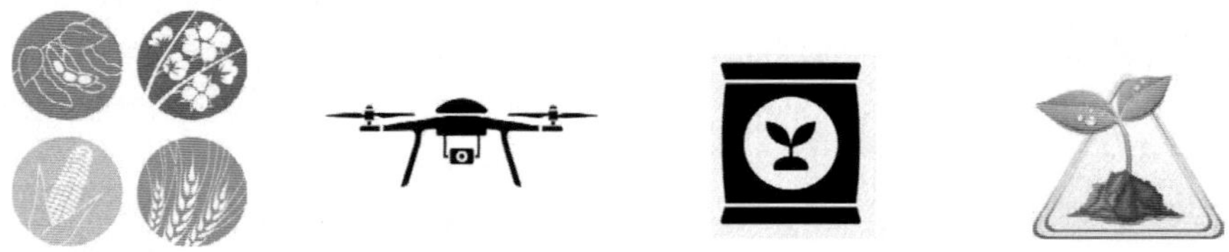

Demand for higher yields and quality	More Information, More sophisticated growing techniques	Farmer and dealer desire for convenience, performance, differentiation	Soil Health, Water quality, Fertilizer efficiency, and Environmental issues

Focus on Agricultural Solutions and Sustainability

Compass Minerals | **Plant Nutrition**

Industry Scorecard

Maximized Knowledge of how to use inputs efficiently:

- **Nitrogen, Phosphorus, Potassium**
- **Lime**
- **Genetics**
- **Crop Protection**

4R nutrient stewardship

SPECIALTY CROP NUTRITION PRODUCTS make current nutritional inputs:

- More available
- More efficient

- Global Population is growing at 4-5% rate
- Yields need to increase by 7-8% annually
- Government agency are predicting 1.5% annual yield increases

Compass Minerals | **Plant Nutrition**

The Value of Specialty Crop Nutrition Products: Bridging the Gap Between Price per Ton and Price Per Acre

Compass Minerals | Plant Nutrition

Figure 1. Value of U.S. Production of Fresh Market and Processing Carrots, 1979-2004

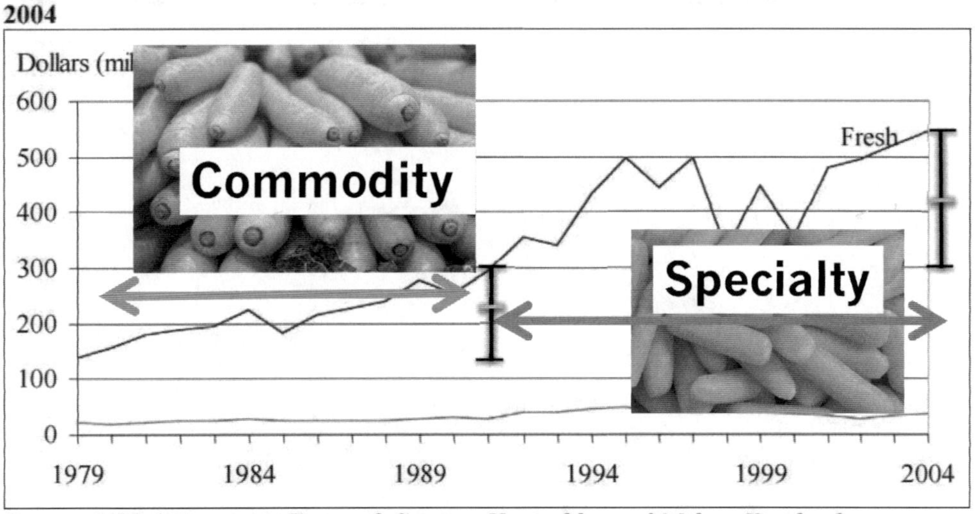

Source: USDA Economic Research Service, Vegetables and Melons Yearbook

Compass Minerals | Plant Nutrition

Truly Novel, Innovative Fertilizer Technology

Investment in fertilizer innovation is limited:

Average New Patent Applications/yr, 1996-2006*	
US Chemical	US Fertilizer
4,500	**8**

We are leading innovation in the industry

- **16** fertilizer patents, 2009 to 2012
- **40%** belong to Compass Minerals.

Source: MarketsandMarkets, 2013, Micronutrient Market – Global Trends and Forecast to 2017

*Table represents the number of patent applications of the U.S. Patent office for the period of 1996 – 2006 in the Fertilizer Manufacturing and Chemical Sectors. Source: Eurostat, R&D Statistics, 2008

Compass Minerals | **Plant Nutrition**

IPNI Data Show Extent of Zinc Deficiencies

Percent of Samples Testing Less Than 1.0 ppm DTPA Equivalent Zn in 2015

- No data available
- 1-19% of soil samples report zinc deficiencies
- 20-34% of soil samples report zinc deficiencies
- 35-50% of soil samples report zinc deficiencies
- 51-65% of soil samples report zinc deficiencies
- 66% or more of soil samples report zinc deficiencies

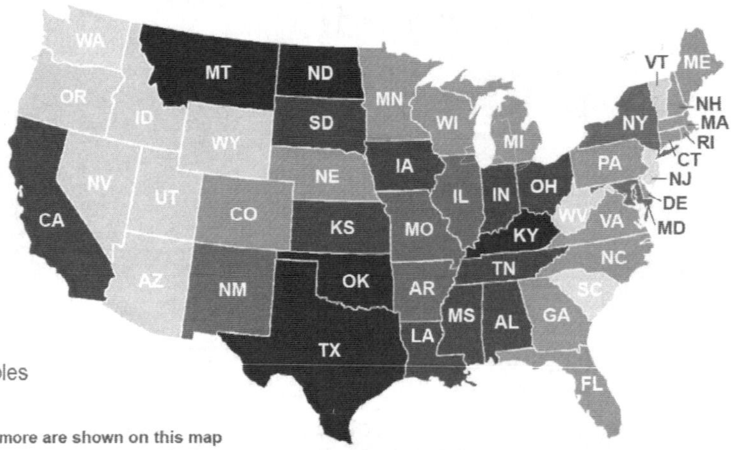

*Only states with 2,000 samples or more are shown on this map

Compass Minerals | **Plant Nutrition**

What's Causing Lower Levels?

Univ. of Illinois:
Hybrids with insecticide traits removed up to 27% more zinc per bushel

		Non-RW Resistant Hybrids	Rootworm Resistant Hybrids	Difference
Yield/Acre		179 bu	205 bu	14%
Nutrient removal / acre	N	110 lbs	126 lbs	14%
	P	17 lbs	21 lbs	24%
	K	26 lbs	31 lbs	19%
	S	8.9 lbs	10.4 lbs	17%
	Zn	2.2 oz	2.8 oz	27%

Source: *Fred Below, Univ. IL*

Compass Minerals | **Plant Nutrition**

High Yields Remove Nutrients

Nutrients with High Percentage GRAIN removal

NITROGEN 58%
PHOSPHORUS 79%
SULFUR 57%
ZINC 62%
BORON 23%

New Hybrids + High Grain Removal = Potential to Mine Soils

Compass Minerals | **Plant Nutrition**

Compass Minerals' Specialty Plant Nutrient Portfolio

wolf-trax
INNOVATIVE NUTRIENTS

Nu-Trax P+™

Micronutrients & Specialty Plant Nutrient Products

- Targets all crops
- Innovative products with patented technology
- Marketed globally

Simple Solution to Old Problems

- Early Season availability key to ROI
- Overcomes SCARACITY of nutrient offered as a granular nutrient
- Overcomes poor distribution of nutrient due to settling of micronutrient within fertilizer blend

Creating New Opportunities

- Differentiate and add value to dry fertilizer
- Technology Platform to make fertilizer more available to this years crop
- Better nutrient utilization by the crop means less waste into the environment

Compass Minerals | Plant Nutrition

Simple Solution: *EvenCoat*™ Technology

- **A more efficient application option**
- **Patented coating technology**
  - Each and every fertilizer granule is coated
  - Improves distribution in the field
  - Use less nutrient to get better uptake

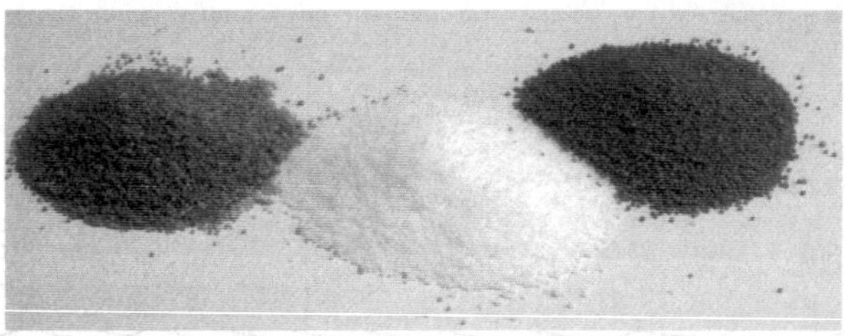

Each and every fertilizer prill is evenly coated with up DDP Nutrient (up to 1% coating per blend).

Compass Minerals | Plant Nutrition

Better Delivery of Nutrients with DDP

Highly Effective
PlantActiv™
formulation
+
Better distribution with
EvenCoat™
Technology
=
Earlier delivery of nutrients

Compass Minerals | Plant Nutrition

Innovation with Zinc Oxide

PlantActiv™ Formulation

Chemical Formulation

- Wolf Trax DDP includes multiple sources of nutrients
  - Dual-Action™ activity
- **Product formulated to minimize soil tie-up**

Compass Minerals | Plant Nutrition

The Science Behind Wolf Trax was Established Years before Wolf Trax Inc.

Rice Response to Granular Zinc Sources Varying in Water-Soluble Zinc

Nathan A. Slaton,* Edward E. Gbur, Jr., Charles E. Wilson, Jr., and Richard J. Norman

Fertilizer Research 33: 249–255, 1992.
© 1992 *Kluwer Academic Publishers. Printed in the Netherlands.* 249

Crop response to level of water-soluble zinc in granular zinc fertilizers

J.J. Mortvedt
National Fertilizer and Environmental Research Center, Tennessee Valley Authority, Muscle Shoals, Alabama 35660, USA

Published September, 1967

DIVISION S-2—SOIL CHEMISTRY

Zinc Movement In Soil From Fertilizer Granules[1]

J. J. Mortvedt and P. M. Giordano[2]

Published January, 1969

Response of Several Corn Hybrids to Level of Water-Soluble Zinc in Fertilizers[1]

P. M. Giordano and J. J. Mortvedt[2]

Reactions of fertilizer zinc in soil and their effect on Zn availability

Fien Degryse, The University of Adelaide

Biol Fertil Soils (2000) 31:343–347 © Springer-Verlag 2000

ORIGINAL PAPER

R. Jay Goos · Brian E. Johnson · Marie Thiollet

A comparison of the availability of three zinc sources to maize (*Zea mays* L.) under greenhouse conditions

Compass Minerals | **Plant Nutrition**

University of Minnesota EXTENSION | NUTRIENT MANAGEMENT

FO-0720-C—(REVISED 2016)

Zinc for Crop Production

Apurba K. Sutradhar: Research Associate
Daniel E. Kaiser, Carl J. Rosen, and John A. Lamb: Extension Specialists in Nutrient Management

"Zinc oxide (78-80% zinc) can correct a zinc deficiency but is slowly soluble and not effective in a granular form. To effectively correct a zinc deficiency, zinc oxide must be finely ground. Spreading any finely ground material is a problem in Minnesota because of the wind. So use of finely ground zinc oxide is limited to situations where suspension fertilizers are used."

Purdue University Department of Agronomy
Soil Fertility Update
June 2012

"Zn oxide is the predominant component it will not be a good source of Zn. Zinc oxide and Zn sucrate are poor sources of Zn when granular because of their low solubility. Water insoluble materials, such as Zn oxide, can provide plant available Zn when powdered but are difficult to spread in that form."

Compass Minerals | **Plant Nutrition**

Zinc Source and Application Type

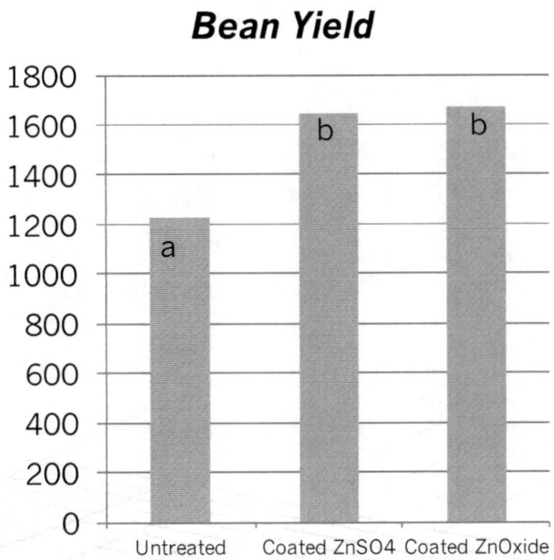

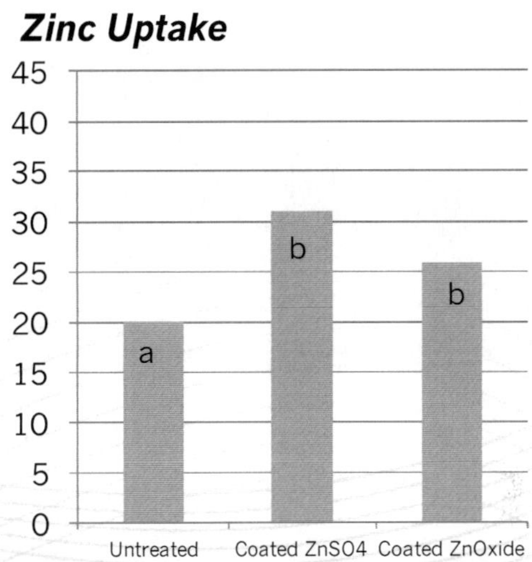

Soil Science Soc . Proc. Pg. 635-636. 1965 Ellis, et.al. Effect of method of incorporation of zinc in fertilizer on zinc uptake and yield of pea bean.

Compass Minerals | Plant Nutrition

Zinc Sulfate – Particle Size Impacts Rates of Effectiveness

Much LOWER rates of a fine powder can perform equal or better than MUCH higher rates of ZnSO4 as a granule

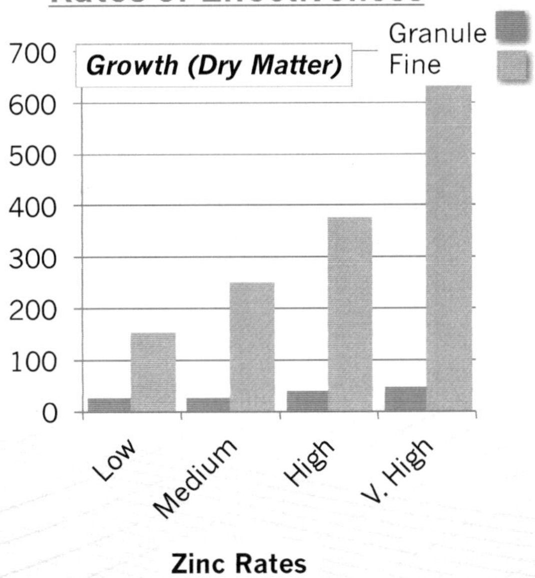

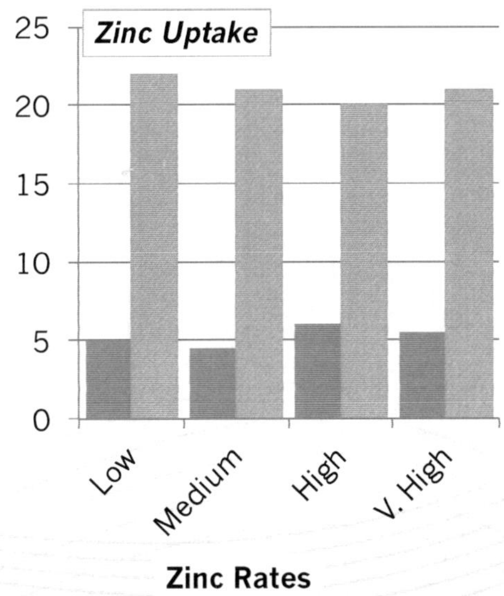

Biology & Fertility of Soils. 2000 31:343-347. Goos R.J., et.al. A comparison of the availability of three zinc sources to maize (Zea mays L.) under greenhouse

Compass Minerals | Plant Nutrition

Zinc Source Matters Less than Particle Size

Products with smaller particle size can be used at much lower rates than granular products IF the nutrient can be evenly delivered to the field.

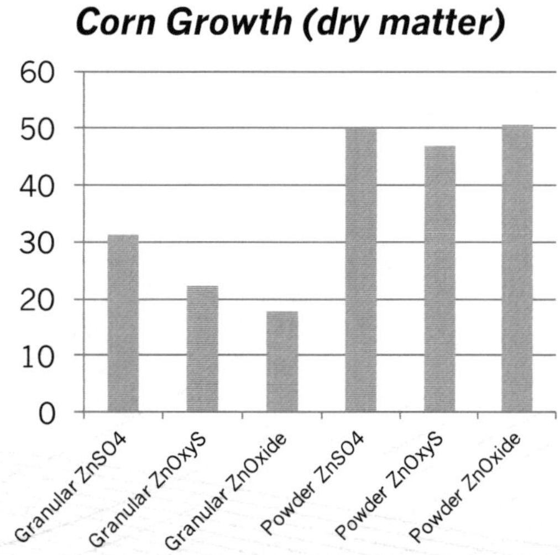

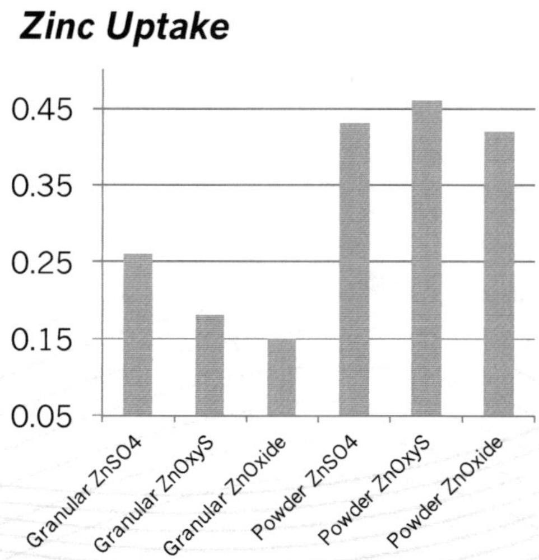

Fertilizer Research 33:249-255, 1992. J.J. Mortvedt Crop response to level of water-soluble zinc in granular zinc fertilizers

Compass Minerals | Plant Nutrition

Zinc Oxysulfate – Same rate, different particle size

Particle size **impacts performance**, smaller particle sizes are more efficient when compared to the same rate as a larger particle.

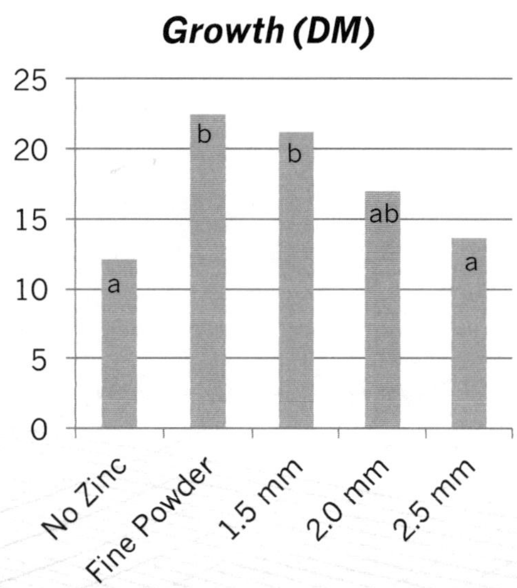

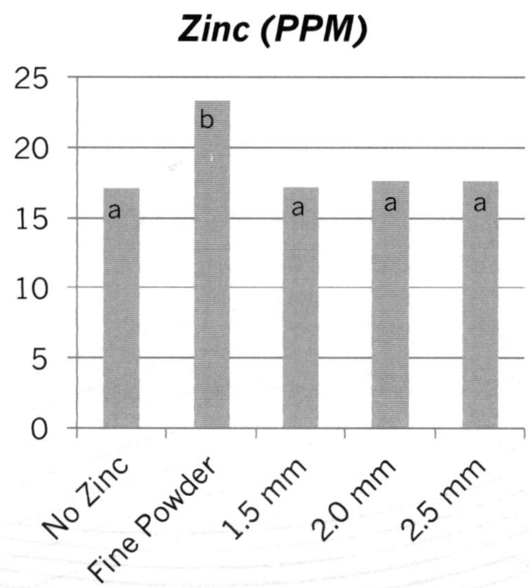

AAES Research Bulletin 963. 2000. Liscano, J.F., et.al. Zinc availability to rice from seven granular fertilizers

Compass Minerals | Plant Nutrition

Sledgehammer . . . To Value

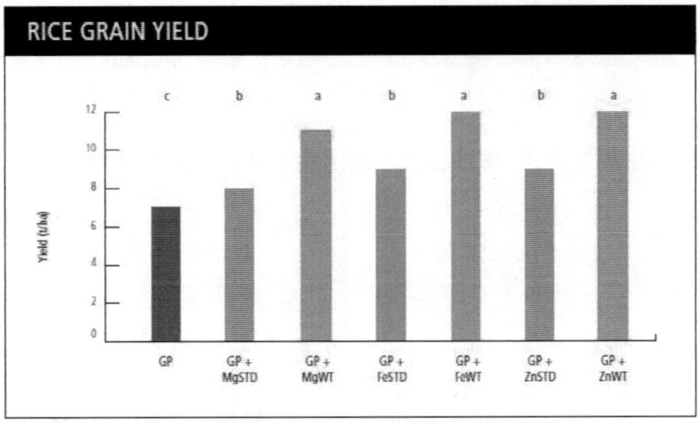

Figure 2 - Effects of Wolf Trax DDP fertilizers on rice grain yield at harvest. Values followed by the same letters do not differ at $P<0.10$.

Percent Reduction in Materials Applied

Magnesium 94%

Iron 95%

Zinc 96%

All Wolf Trax DDP treatments were statistically significant when compared to granular version

Compass Minerals | **Plant Nutrition**

Summary

- Zinc Oxide can play a role in Specialty Nutrition
- Focusing on efficiency and performance can provide value in the marketplace
- The soil zinc market should grow, driven by:
  - Higher crop removal
  - Decreasing soil organic matter levels
- Zinc is part of the "Big 5" and needs to be managed

Compass Minerals | **Plant Nutrition**

Thank You

Zinc Oxide: The New Norm

February 2017

DUST
METAL
OXIDE
FINES

U.S. ZINC®

What's going on?

ZnO is not usually a headline-making industry ….

- ZnO considered a static market
- Historic slow (GDP) growth industry

And now we are seeing …

- Changes in the competitive landscape
- International trade uncertainty
- Globally instituted regulations
- Economic regionalization

➢ There has been a ***pivot*** in the industry … so what does this New Norm mean for us as a supplier, as an industry?

U.S. ZINC®

U.S. zinc overview

- Global provider of four product lines, serving broad array of markets:
  - **Zinc Oxide** - tire, rubber, chemicals, plastics, pharma
  - **Zinc Dust** - paint & coatings, mining
  - **Zinc Metal** - galvanizing
  - **Zinc Fines -** agriculture, smelting
- Founded in Houston, Texas 68 years ago
- Full global integration with affiliation with Votorantim, one of the five largest metal producers
- Largest zinc oxide producer in the world with 150,000 mt/yr
- One of the largest zinc metallics recycler in the world
- 6 Manufacturing facilities: Dust & Metal: Houston, Texas
  - Oxide: Brazil, China, & US (Millington & Clarksville, TN)

U.S. ZINC®

A recycler

OUR CYCLE IS RECYCLE®

- Outlet for hot dip galvanizers' residues:
  - Skims to produce Prime Western Zinc Metal
  - Bottom Dross to produce zinc dust

- Secure outlet for continuous galvanizers' residues:
  - Top Dross to produce zinc oxide

U.S. ZINC®

Macro market and ZnO

We are seeing interesting times …

- US and global economy continues its gradual steady upward movement
  - US GDP growth ⬆ EU GDP growth ⬆ Japanese GDP ⬆
  - Steel companies increasing output in North America (Chinese import tariffs)
  - Galvanizing to follow … ?
  - Oil field lowered capacity utilization, but is starting to increase
  - Infrastructure spending - $3B infrastructure bill passed last year, further spending being contemplated now in Congress
- Auto and Tire industries are at historical highs in the U.S.
- LME is at a 9-year high, off a 9-year low
- Nationalism rhetoric
- Political uncertainty
- Global trade may be redefined with trade agreements and tariffs in flux

U.S. ZINC®

Auto and tire market and their effects on oxide

- Auto industry continues strength, even if seeing a plateau
  - ~18M vehicles sold in 2016; 17.8M in 2015
  - SAAR Jan 17.6M – a strong start
  - Auto sales has been a metric of overall economy health

Record Number of Miles Driven in U.S. Last Year

Americans drivers put 3.22 trillion miles on the nation's roads last year, up by 2.8 percent from 2015. A better economy and lower gas prices fueled the increase, which highlights infrastructure needs.

Source: http://www.npr.org
http://www.npr.org/tags/516519925/miles-driven

U.S. ZINC®

Auto and tire market and their effects on oxide

- Chinese tire import tariffs lapse in 2012 - tire imports in to US reached ~ 65m by 2013
  - Import tariffs reinstated 2014
  - Tire imports from China decrease to under 30M by YE 2016
  - Tire production moves to SE Asia and US
    - ➢ Leading to a global shift in global tire production
- Tire companies continue to expand
  - More than ten major tire producers have announced, are currently undergoing or have completed expansions or are building new production facilities:
    - ➢ Bridgestone, Conti, Giti, Goodyear, Hankook, Kumho, Michelin, Pirelli, Sentury, Toyo, Yokohama
  - Estimates put additional US tire capacity 2014 through 2021 at over 45M tires

U.S.ZINC®

Supply chain and the ZnO market

Logistics in 2016 presented challenges due to national and global economic environment:

- **US market**
  - Saw an overcapacity of trucks, lowering prices YoY
  - Incoterms shift from CPT to EXW due to customers demanding control of their supply chain and better rates

- **Export market**
  - Bankrupt companies due to low pricing, leading to consolidation and price increases (Hanjin)
  - Consolidation of shipping companies due to lower profitability
  - Changes to SOLAS regarding safety of life at sea, weight verification and additional controls at US ports (capsizing of vessels due to weight)
  - Pricing fluctuation; but all indications pricing will increase

CONFIDENTIAL AND PROPRIETARY INFORMATION - This document has been developed for the exclusive internal use of U.S. Zinc Corporation. Its receipt or possession does not convey any rights to disclose its contents. Reproduction or further distribution of this document without the permission of U.S. Zinc Corporation is strictly prohibited.

U.S.ZINC®

Supply chain going forward

The effects of an improving market and regulatory changes will be felt:

- **US market**
  - With economic improvement will come less supply of trucks in 2nd half of 2017
  - Trucking industry adjustment to new DOT regulations, in particular the electronic logging that will control hours of operation for all trucks
  - Possible regulation changes in US to Zinc Oxide classification with the DOT
- **Export market**
  - Continued consolidation of shipping companies
  - Asia-Europe and trans-Pacific shipping contract rates are set to rise through 2017 as the higher cost of fuel is strengthening ocean carriers' hands; even though, the current environment is of overcapacity.
  - Source: Bamard, Bruce "Contract rates forecastto rise in 2017 for first time since 2010," www.joc.com, 12/15/16
- **Procurement**
  - Costs increases from GHS changes and new labeling
    - New designs, additional markings and tags to comply with new GHS rules and regulations
    - Non Regulated in US but environmental hazardous in EU and other countries.

U.S. ZINC®

Supply Chain: ZnO import duty tariffs

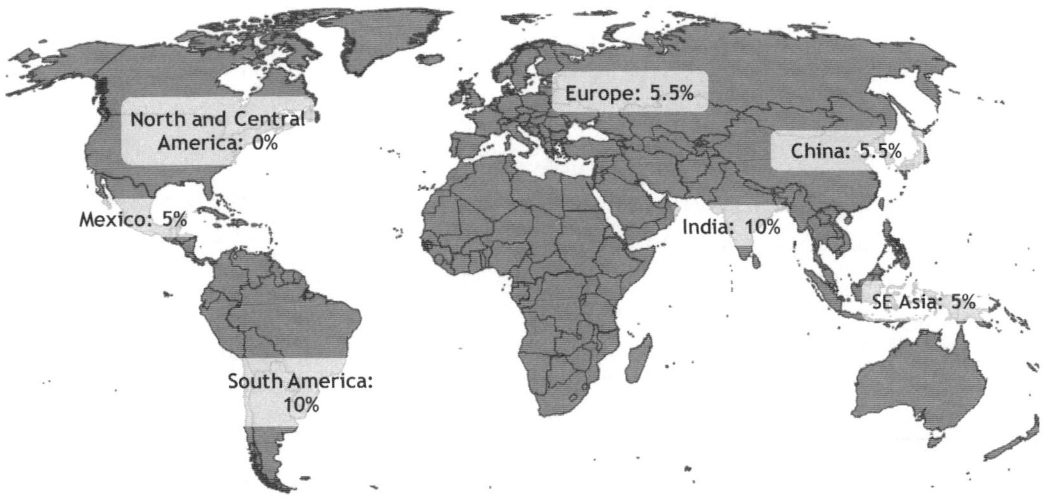

- In and out of US
- **Will these change under the new government?**

U.S. ZINC®

The regionalization of the zinc oxide market

… Yet we still see the regionalization and localization of markets from an economic standpoint

5 Year History of N. American ZnO market

- Largest N. American facility shut down
- Expansions across competitive landscape
  - Led to oversupply

- Fear of tariffs may put customers at risk if they are importing large volumes from other countries
- Regulations of secondary raw materials as well as finished good could change landscape quickly
  - On one hand, globalization with regulations, becoming consistent
  - On other hand, from economic standpoint, regionalization

Going Forward: How to manage the new norm

What does this mean for the oxide market?

? How can we capitalize on this

? Challenges and Opportunities

- ✓ Must be ready to react to changes quickly
- ✓ Suppliers purchasing more regionally?
- ✓ Competing in unfavorable currency environment
- ✓ Customers investing more in US due to potential tax law changes
- ✓ Duties to other countries may change to make North America more competitive

THE NEW NORMAL IS CHANGE

CONFIDENTIAL AND PROPRIETARY INFORMATION - This document has been developed for the exclusive internal use of U.S. Zinc Corporation. Its receipt or possession does not convey any rights to disclose its contents. Reproduction or further distribution of this document without the permission of U.S. Zinc Corporation is strictly prohibited.

U.S. ZINC®

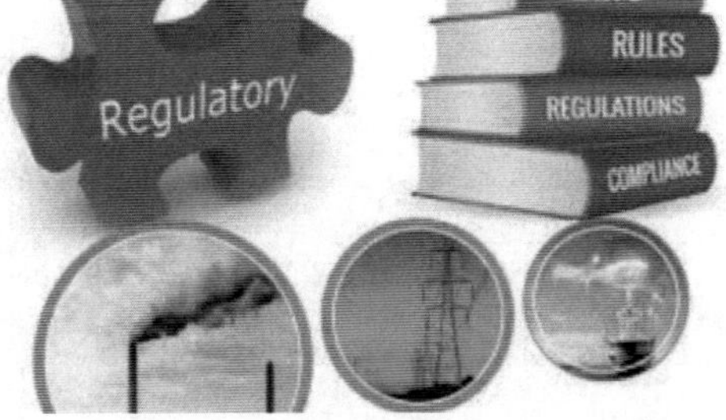

Update: EU Regulatory Issues and beyond

Christine Spirlet
Manager, Regulatory Affairs

Overview

- Introduction: the regulatory effect
- REACH - **R**egistration, **E**valuation, **A**uthorization and Restriction of **Ch**emicals
- Nano regulatory update
- Veterinary medicine, animal feeds, oral care
- Global issues
- Conclusions

The "regulatory effect"

Chemical management workload is still growing

Chemical Watch January 2017 survey says…

Companies will increase staff working on chemical compliance

Or increase use of external service providers **(IZA)**

Timing:

next 12 months => 5 years ahead

Teams getting larger, job security and prospects solid, and use of external service providers increasing.

Main Regulatory Drivers

Europe ⇧

- REACH 84%
  - 2018 registrations 69%
  - SVHC obligations 49%
  - Evaluation & updates 41% (NEW)
- CLP 40%

USA ⇧

- US EPA workplan 35%
- CA Safer Consumer Products 21%

Asia ⇩

- China 35%
- South Korea 26%
- Japan 16%
- Taiwan 14%

% = mentioned by survey participant

REACH update for ZnO

REACH dossier background

1st registration in 2010 (115 legal entities registered to date)

2012 update (new GHS/CLP environmental classification criteria)

2013 update (Sewage Treatment Plant exposure scenario)

2015 update (Occupational Exposure Limit revision & scientific review nano form)

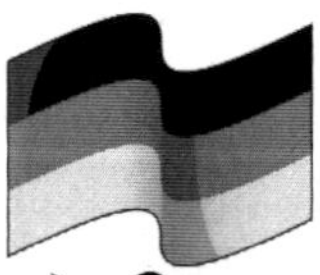

REACH ZnO dossier evaluation

- Germany proposed ZnO for **CoRAP list** (**Co**mmunity **R**olling **A**ction **P**lan) for evaluation in 2017
- Focus is on nanoform with following rationale (based on the 2013 registration dossier):

No allocation for a nanoform to a specific use.

Lack of adequate exposure assessment is a major concern.

Material characterization is insufficient

Toxicity endpoints information missing

Wide dispersive uses not documented

IZA work: 2015 & 2016 updates

- Scientific revision of nano data : recent literature review
- Compilation of exposure data from manufacturers

Specific use description	Exposure assessment and risk characterization	Specific physico-chemical characterization
Toxicity endpoints documentation	Environmental assessment	

Evaluation process

German Authorities sent a questionnaire to be filled by companies to document individual data: February 24th

↓

Formal evaluation will commence in March

New exposure limit "soon"

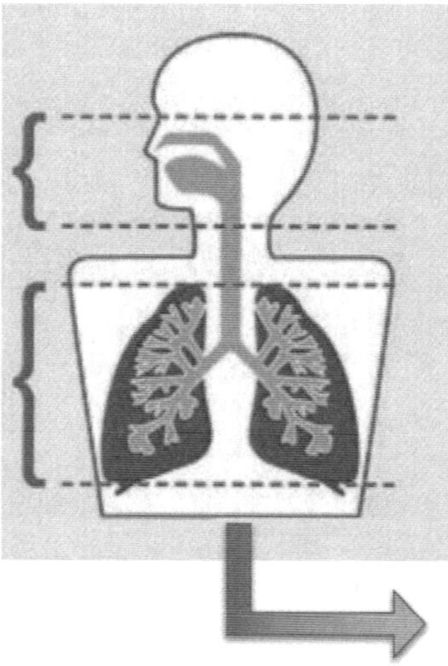

Introduction of a local inhalation DNEL for the respirable fraction of **0.5 mg ZnO/m³ or 0.4 mg Zn/m³ based on rat study**

Zinc REACH consortium co-financed human volunteer study on ZnO to balance with animal data

Discussions ongoing:
NOAEL at 0.5 or 1 mg ZnO/m³?
AF 2 (4h exposure time)?

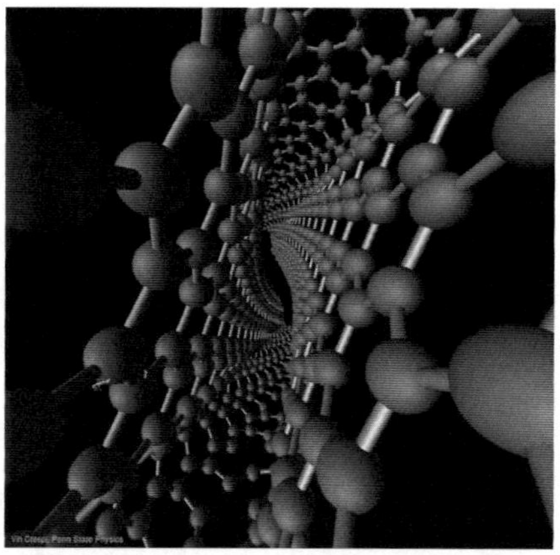

Nano regulatory update

Revision of recommendation

Definition	• EC Recommendation on a definition of nanomaterial, published in 2011 (EC Recommendation 2011/696/EU).
Revision of recommendation	• Objective is not to change scope but clarify terms and facilitate implementation
Main elements of change	• Particles clarification • Removal flexibility of threshold • Removal VSSA as method to determine if nano but include to demonstrate not nano

Guidance for implementing definition

Roadmap (tentative)	
Draft revised definition	Nov/Dec 2016
Public consultation	Nov/Dec 2016 – Jan/Feb 2017
Revised definition – adoption	Feb/Mar 2017
JRC Report 1 (tech report supporting implementation of definition)	Adoption of the revised definition
JRC Report 2	End 2017

No EU Nanomaterials Registry but an EU Nanomaterials Observatory

The **aim** of the Observatory is to increase the transparency on nanomaterials without creating new obligations for companies. **Target**: from MSCAs 'down to' workers and consumers.

Project will be delegated to ECHA.

- will manage the Observatory, taking over from information gathered by JRC.
- will collect available or voluntarily provided information, filter it, and post it.

ZnO in Veterinary Medicine

Background

Nov. 2016: Referral Assessment Report showing that environmental risk outweighed the benefits of using ZnO.

Early Dec. 2016, CVMP Opinion released, advising ban ZnO in veterinary medicine immediately

Process

Dec. 2016 & Jan 2017: Requests for re-examination of the CVMP Opinion were sent to the EMA (European Medicines Agency)

Jan 2017: EU COM Standing Committee say withdrawal should be delayed

Member States to submit an impact assessment by **March 13th** (pig health, costs, alternatives, delay)

Process

Early February: MAH submitted detailed grounds for re-examination to the EMA

→ **Feb. 14th** oral explanation (IZA as environmental risk assessment expert).

→ **February 24th** Rapporteurs circulate a revised assessment report (still negative)

→ **March 1st** MAH (+IZA) to comment on the revised assessment report

→ **March 7th** Final CVMP Opinion circulated to CVMP members

→ **March 13th** MS to submit their impact assessment

→ **March 16th** Adoption of the final Opinion

What if?

EMA opinion is reversed and positive

- E.U. Commission and Standing Committee on Agriculture/Medicines
- MS pro or against ZnO

EMA opinion is not-reversed and negative

- E.U. Commission and Standing Committee have already been alerted that a sudden ban will cause devastating problems regarding pig production.

2017 IZA Conference - Palm Springs (CA) International Zinc Association

Next steps

EMA/CVMP

- IZA and other consultants will write comments on the CVMP's revised assessment report
- These comments will have to be sent to the EMA directly (by the 1st of March).
- If proven that the rapporteurs misinterpreted our environmental assessment; there is a possibility that the revised assessment report will have to be corrected...

Commission and Member States (MSs)

- UK NPA and PVS have submitted reports to the COM
- EU Groups such as
  - Copa Cogeca (EU farmers and agri-cooperatives)
  - Fefac (EU Feed Manuf. Federation)
  - Fefana (EU Association of Specialty Feed Ingredients and their Mixtures).
  - The farmers lobby (UK) is also very strong (ban will directly influence their profitability
- Investigate which MS is for or against (lobbying activities)

ZnO in Animal Feeds

Reviewed authorization

COMMISSION IMPLEMENTING REGULATION (EU) 2016/1095

of 6 July 2016

concerning the authorisation of Zinc acetate dihydrate, Zinc chloride anhydrous, Zinc oxide, Zinc sulphate heptahydrate, Zinc sulphate monohydrate, Zinc chelate of amino acids hydrate, Zinc chelate of protein hydrolysates, Zinc chelate of glycine hydrate (solid) and Zinc chelate of glycine hydrate (liquid) as feed additives for all animal species and amending Regulations (EC) No 1334/2003, (EC) No 479/2006, (EU) No 335/2010 and Implementing Regulations (EU) No 991/2012 and (EU) No 636/2013

(Text with EEA relevance)

Applied limits for ZnO

Species or category of animal	Content of element (Zn) in mg/kg of complete feed with a moisture content of 12 %	Other provisions	End of period of authorisation
All animal species No maximum age	Dogs and cats: **200** (total)	The additive shall be incorporated into feed in the form of a premixture.	27 July 2026
	Salmonids and milk replacers for calves: **180** (total)		
	Piglets, sows, rabbits and all fish other than salmonids: **150** (total)		
	Other species and categories: **120** (total)		

ZnO in Oral Care

DG Internal Market, Industry, Entrepreneurship and SMEs

Request for a scientific opinion (SCCS):
Submission I on water-soluble Zinc salts used in oral products

Deadline: March 2017

Health assessment of zinc salts in oral hygiene products and toothpaste. BfR July 2014

Consumer safety and exposure assessment for zinc in oral care products. Cosmetics Europe. February 2016

Background

Currently allowed for use in cosmetic products in concentrations of up to 1% as Zinc with a wide range of functions (2009) → Tolerable total uptake level (UL) for Zinc up to 25 mg (2003 SC Food)

→ BfR concluded 1% Zn in toothpaste and mouthwash can lead to excess Zn intake (focus on children and teens) → Feb. 2016, Cosmetics Europe: aggregate exposure assessment for children and adults: 1% toothpaste, 0.1 % for mouthwash safe

Waiting for opinion (March 2017)

Zinc

Water-soluble Zinc salts safe for all age groups at the current allowed concentration of up to **1%** as Zinc when used in toothpastes and at the use level of up to **0.1%** as Zinc when used in mouthwashes?

Alternatively, consider water-soluble Zinc salts when used in toothpastes and mouthwashes safe for all age groups at the lower use concentration of up to 0.1% as Zinc?

Any concerns related to the use of water-soluble Zinc salts in oral products for the particular age group 1 to 18-year old?

Global issues

2017 IZA Conference - Palm Springs (CA)

International Zinc Association

Japan Activity - Restoring the "H" in GHS -

- Inconsistencies between Japan and Europe

GHS Classification	Japan	Europe
Mutagenicity Repro-toxicity STOT	Category 1 & 2 (Potential for Sunset)	Not Classified

- Coordinating with IZA network: JMIA, JLZDC, JICIA, Mitsui, C-P, AIST, RISS, MOE
- IZA working with Icarus Japan to identify mechanism for reclassification

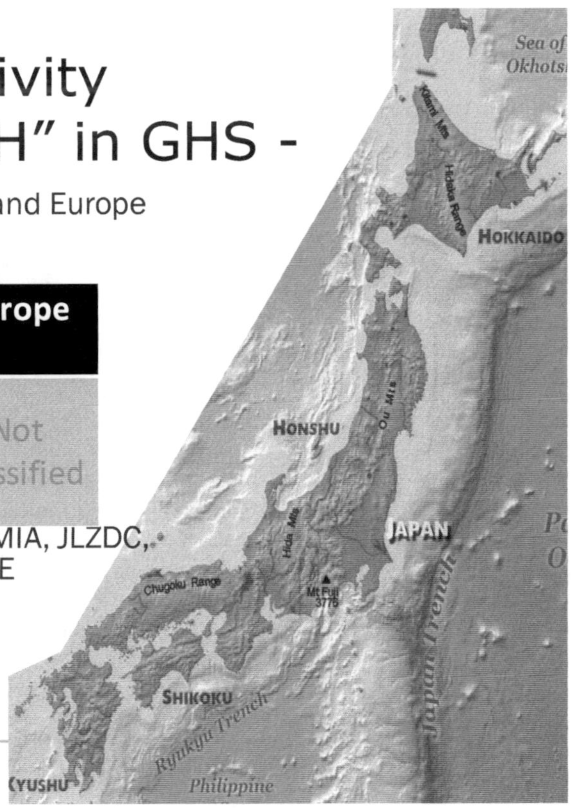

2017 IZA Conference - Palm Springs (CA)

Zn GHS Classification Timeline in Japan

Year	"Classified" Repro-toxic	"Not Classified"
2006	$ZnCl_2$, $ZnSO_4$	ZnO
2009	$ZnCl_2$, $ZnSO_4$, ZnO	-
2013-Current	$ZnSO_4$, ZnO	$ZnCl_2$
Europe (Current)	-	$ZnCl_2$, $ZnSO_4$, ZnO

Consequences:

- Market sensitivities to consumer product use
- Migration to other Asian economies (China, Korea, Thailand, Vietnam, etc.)

Zn GHS Classification migration

http://www.ajcsd.org/chrip_search/sltLst

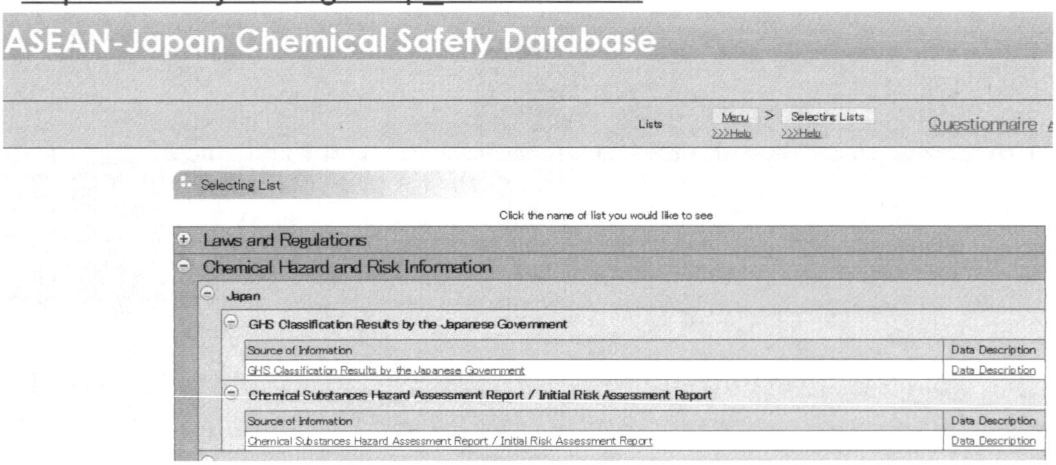

Conclusions

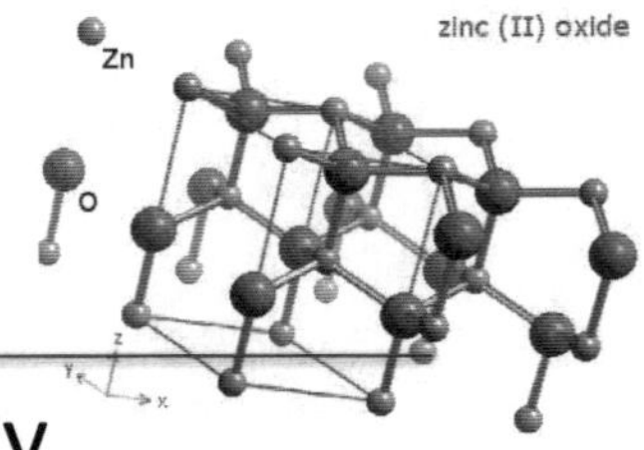

REACH dossier evaluation: ready

OEL revision: provisional value

Veterinary Medicine: uncertain outcome

Oral Care: expecting scientific opinion

Japan classification as reprotox => ?

Thank you!

New Technology Developments in Zinc Oxide

Peter Robinson
Principal
IRM

Shahrom Mahmud
Universiti Sains
Malaysia

International Zinc Association

Zinc and Zinc Oxide Conference

February 26 – March 1 2017

ZINC OXIDE CONSUMPTION BY INDUSTRY

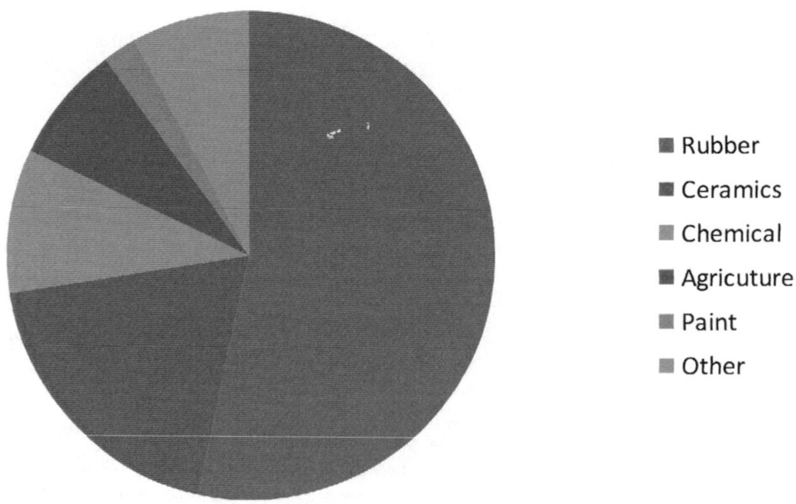

DRIVERS FOR MARKET CHANGE

- SUSTAINABILITY - Availability
- TECHNOLOGY
- ECONOMICS
- DISRUPTION

EFFECT OF NEW TECHNOLOGY ON THE ZINC OXIDE INDUSTRY IN-

Zinc Oxide production

Existing markets

New markets

MACRO versus MICRO

- MACRO INDUSTRY

Building, Automotive, Agriculture, Coatings

Energy

Volume - Commodities

MICRO INDUSTRY

Semiconductors, Solar cells, Sensors, Electronics

Value - Specialities

Zinc Oxide Research

- US Patent applications – last 250 that include ZnO in the claims - a guidepost for technology development - by subject

• Products using ZnO -	59
• Electronic applications -	51
• Thin film, optoelectronics -	47
• Nano -	24
• ZnO particle properties -	23
• Conductive/Varistors/Catalysts -	12
• ZnO manufacture -	3
• Others -	21
	250
• Applicable to volume industries -	4

Conclusion - patents

- Almost all of the patents and hence research work are directed toward uses of zinc oxide which are in Industries that use microscopic quantities of ZnO.

Manufacture of Zinc Oxide – The Metsol Process

- The Metsol Process uses ammonia based leaching which is very selective towards zinc.
- The process is particularly suitable for oxidic resources which have modest zinc assays and are not candidates for conventional zinc recovery.
- Metsol ZnO is white and has very similar characteristics to French Process ZnO
- The Process bypasses the conventional route to ZnO pigment-

METSOL PROCESS & FRENCH PROCESS

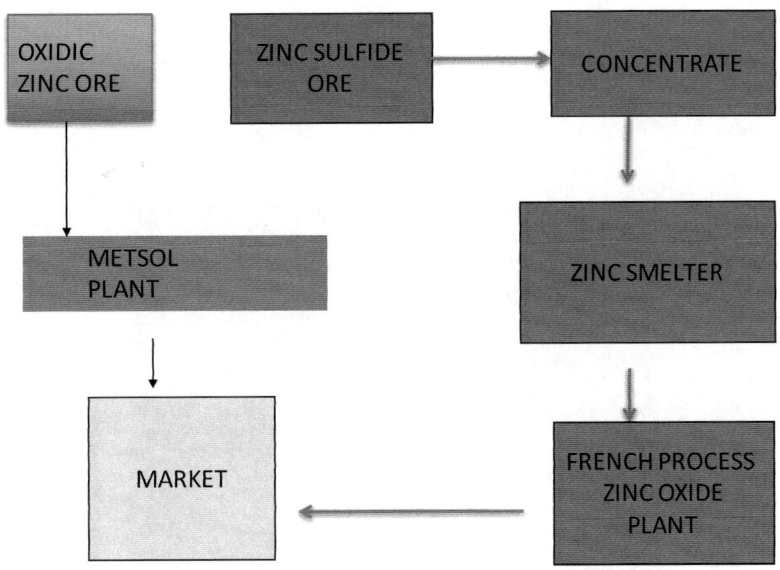

Selected Examples

- Rubber – Anything affecting rubber affects zinc oxide. Areas of concern for tires are run-off into the marine environment and the high density of zinc oxide.
- The 2016 California initiative to reduce the level of zinc in tires – Is this the thin end of the wedge? Could this be an opportunity for nano ZnO?
- Changes in the natural rubber polymer, eg Guayale, Russian dandelion and more recently biomass isoprene.
- Research work at the University of Minnesota has developed a route to isoprene via sugar compounds from biomass such as corn, grass and trees. This apparently can be achieved without solvent extraction

Lithium ion batteries

- Apparently not a conflagration, just a thermal runaway-

Zinc Oxide in Li ion batteries

- Transition metal oxides have gained more and more interest as electrode materials in lithium-ion batteries in the last decades because of their higher theoretical capacity and safety compared with the conventional carbon materials [1–4]. Among them, ZnO, a wide-bandgap semiconductor of 3.37 eV at room temperature with large exciton binding energy of 60 meV, is a multifunctional material for a variety of practical applications such as piezoelectric devices [5], field emission [6], gas sensors [7], dye-sensitized solar cells [8], and photocatalysts [9] due to its excellent physical and chemical properties. ZnO has superior advantages such as low cost, facile preparation, morphologic diversity, and high chemical stability. ZnO nanocrystals with different morphologies have some special properties of physics, so intense interests have been devoted to the synthesis of ZnO with various morphologies such as nanowires [10], nanobelts [11], nanorods [12], and nanosheets [13].
- https://www.hindawi.com/journals/jnm/2016/8056302/

....continued - ZnO to the rescue

- The zinc oxide was synthesised from zinc stearate
- The ZnO in nano form was able to control current flow over multiple cycles
- This is one of a number of initiatives to improve the safety of lithium ion batteries.

Aerogels

- Zinc oxide and other materials, notably silica, can be manufactured with surface areas almost two orders of magnitude greater than French Process. Surface area >250 m2g-1, bulk density 0.04g/ml.
- A relatively costly process but may have uses other than as a laboratory curiosity.

Metal Organic Frameworks

- Metal organic frameworks (MOFs) are structures made up of inorganic nodes, and organic linkers. They contain potential voids which can be used for gas storage and separation, catalysis or drug delivery as well as electronic and battery applications.
- Zinc and copper are the two most common metals used. Many options exist for the organic links. Hydrogen storage was the original envisaged use, however the ability to design particular MOF's is leading to the wider range of applications.

MOF structure

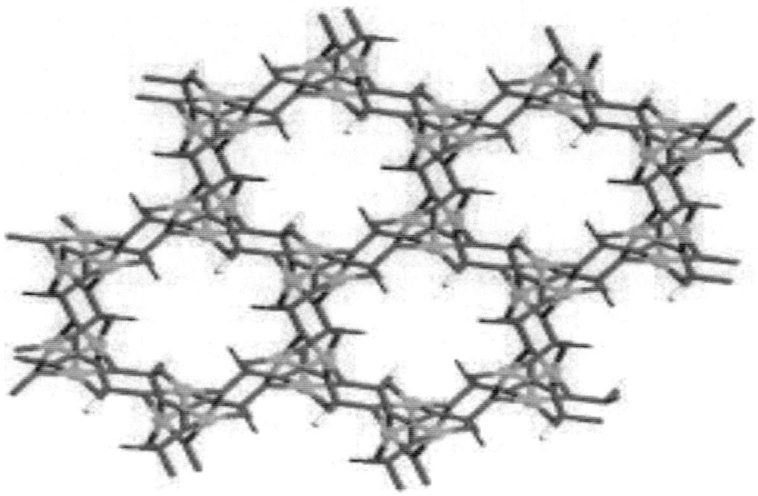

Zinc Oxide in Cancer Treatment

- Zinc oxide has been found to be an effective topical treatment for some cancers.
- It is toxic to cancers but much less so to normal tissue.
- Cytotoxicity has been found to be influenced by the zinc oxide particle size and shape.
- Cytotoxicity is caused by release of ROS, reactive oxygen species and zinc ions
- Work is continuing with delivery methods to concealed cancers and also photochemically activated treatments.

Zinc Oxide in Concrete

- Concrete is the building material of choice – The world market was 4.6 billion tonnes in 2015
- Concrete Industry is under pressure to reduce carbon dioxide emissions.
- IZA work has shown that addition of ZnO to concrete improves the corrosion resistance of rebar and also gives biocidal protection.
- The chemistry of concrete is still not well understood,.
- The chemistry of zinc in concrete is even less understood. The interaction with calcium silicate and calcium aluminate are important. Concretes that cure too quickly appear to be good candidates. An interesting possibility is limestone cement, Contempra.
- In some formulations, high strength concrete has been achieved with over 20% zinc , A chemical that rare in nature, gahnite, zinc aluminate, may be the key, The Mho hardness is 8.

THE MACRO MARKET - CONCRETE

- We have pursued niche markets in concrete for example in retardation, grouts and additives as well as alternative cementitious materials such as GGBFS.
- More research work on zinc oxide in concrete is being published although it is usually related to the clinker precursor.
- CONTEMPRA PLC cement has added ground limestone which saves on cost (&CO2 emissions). It also has high reactivity and hence may be the vehicle to introduce the beneficial ZnO properties.
- Concrete represents the major opportunity for zinc oxide.

LIGHT AT THE END OF THE TUNNEL?

- Batteries for load levelling and other large scale energy storage systems need to be of low cost sustainable materials, capable of multiple cycles.
- Examples of such batteries include the Zn.Fe and Zn.Ni, as described earlier in the conference.
- As the zinc needs to be regenerated in a reactive form, the starting material is zinc oxide.
- Solar panels systems and wind generation facilities require load levelling A single large 2.4 MW wind turbine could use 8 tonnes of ZnO in the case of Zn.Fe batteries. A large scale wind energy project such as the one planned off Hornsea in the North Sea could require 4,000 tonnes of ZnO.
- Clearly a Macro opportunity for ZnO.

The Green Fields of .....ZnO

Spring ZnO Flowers

Thanks

Any Questions?

PRIVATE AND CONFIDENTIAL

INDIAN ZINC OXIDE INDUSTRY AN OVERVIEW

ANIRUDH JHUNJHUNWALA
CEO
J.G. CHEMICALS PRIVATE LIMITED

INTERNATIONAL ZINC OXIDE CONFERENCE
FEBRUARY 2017

A BDJ GROUP ENTERPRISE

Table of Contents

A BDJ GROUP ENTERPRISE

Our Legacy: Enterprises Based on Ethics and Principles

JG CHEMICALS

Our Founder

Late Brahma Dutt Jhunjhunwala

Industrialist. Philanthropist. Freedom Fighter

Snapshot

- **History:** The BDJ Group was established in the early parts of 1900 in Kolkata (Eastern India)
- **Presence:** Offices and manufacturing locations located across India
- **Size:** Employs over 3,500 people across India
- **Sectors:** The founders initially started off with trading in agricultural commodities and then expanded across glassware, refractories, mining, chemicals, real estate and financial services industries
- **Management:** 3rd generation family owned and professionally managed group

Leadership Position Across a Wide Range of Sectors

JG CHEMICALS

Sectors		Description	Companies
Zinc Oxide		▪ India's largest french process zinc oxide producer with multi location production facilities.	▪ JG Chemicals Private Limited ▪ BDJ Oxides Private Limited
Glassware	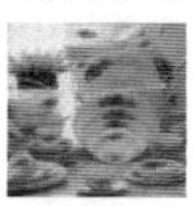	▪ India's number one and only manufacturer of opal and crystal glassware; also present in lower end glassware products with over 50% market share	▪ La Opala RG Limited ▪ Balaji Glass Industries Private Limited
Electrical Components		▪ Manufacturers of private label electrical components as well contract manufacturer for leading European light fittings manufacturer	▪ Radha Electricals Private Limited
Trading		▪ Trading in ferrous / non-ferrous metals and products for rubber & tyre sector with a network of sourcing relationships across India and abroad	▪ BDJ Impex Private Limited
Financial Services and Real Estate	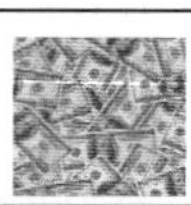	▪ Mid tier wealth management and advisory business; also involved in real estate investment groups	▪ Jayanti Commercial Limited ▪ Vision Projects and Finvest Private Limited

BDJ Group has established itself across a wide range of sectors; leadership, trust and commitment form the backbone of every group enterprise

JGC: A Snapshot of India's Leading ZnO Producer

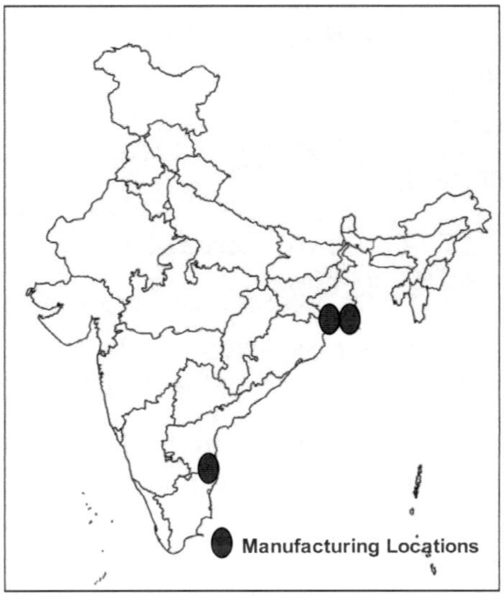

Key Facts

- **Established in 1975**
- **Largest ZnO Manufacturer in India**
- **3 Production Facilities**
- **Capacity**
  - **Installed:** 30,000 MTPA
  - **Projected:** 60,000 MTPA by 2022
- **ISO 9001 (2015) Certified and REACH Registered**
- **Diverse Customer Base**
- **Largest Importer of Zinc Scrap in India**

JGC is India's largest and most reputable zinc oxide producer with a diverse customer base and global sourcing arrangements; it enjoys the "preferred supplier" status amongst its leading tyre customers

A BDJ GROUP ENTERPRISE

5

JG Chemicals Private Limited: India's Leading Zinc Oxide Producer

New plant of JGC has the state of the art infrastructure being designed as per International Standards and Manufacturing Practices; it is spacious and with modern plant machinery

BDJ GROUP

6

Way Forward: We at JGC…Are Open To…

JG CHEMICALS

- ☑ **Information Sharing:** Sharing of information within industry participants for the benefit of the industry
- ☑ **Best Practices Sharing:** Sharing and transfer of best practices across participants in the zinc oxide industry
- ☑ **Technological Tie-ups:** Technological tie-ups with international producers in the area of zinc oxide and other products
- ☑ **Joint Ventures:** Establish joint venture businesses in India and abroad with international companies in the field of zinc oxide and other industries
- ☑ **Utilize Local Expertise:** Leverage our strengths and reach in the Indian market together with our partners

We are open to collaborating with international companies to establish, expand and create the market leader of tomorrow in zinc oxide and other industries

A BDJ GROUP ENTERPRISE

7

Table of Contents

JG CHEMICALS

A BDJ GROUP ENTERPRISE

8

India: Growth Levers in Place for the "High Growth Decade"

JG CHEMICALS

Issue	Description
Increasing Domestic Demand	▪ With a population of over 1.25bn and growing, there is a strong domestic demand which is driving consumption; rising income levels further fuelling the demand
Availability of Skilled Man-power	▪ India has a large pool of skilled man-power which will help attract global and Indian companies to invest in manufacturing capabilities in India
Stable Political Environment	▪ With the current BJP regime, there has been an era of political peace and a strong pro-industry move. This has helped attract global corporations to India
Increasing FDI	▪ The increasing presence of global corporations in India has led to a rapid rise of FDI to over US$36bn in the last 12 months; the highest ever in its history
Demonitisation	▪ A strong move by the Indian government to bank old Rs 500 and Rs 1,000 notes has led to an increased push towards a cash less economy; the drive has helped reduce corruption and will increase transparency

Per Capita Consumption

	India	China	World
Steel	60	488	222
Rubber	1.16	5.80	3.12
Auto	22	183	145
Cement	225	1581	500

Steel, Rubber, Cement – Kg per person
Automobile: No of vehicles per 1000 persons

India's economy is well positioned to grow over the next decade due to a strong domestic demand; global factors will have a lesser impact on India due to its strong domestic economy

Source: Bloomberg, CIA Fact Book, World Bank, IMF, Ministry of Finance - India

India: Key Economic Indicators Project High Growth

JG CHEMICALS

Consistent Increase in GDP Driven by Strong Domestic Demand and Make in India Campaign

Indian GDP Data in US$ bn from 2006 to 2020 (Projected)

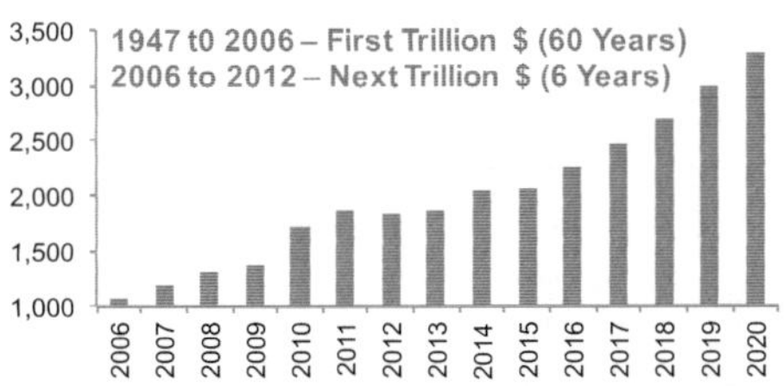

India's Share in Global GDP Will Increase to Over 3.5% from the Current 3% by 2010

India and World GDP Growth Rate and India's Contribution to Global GDP

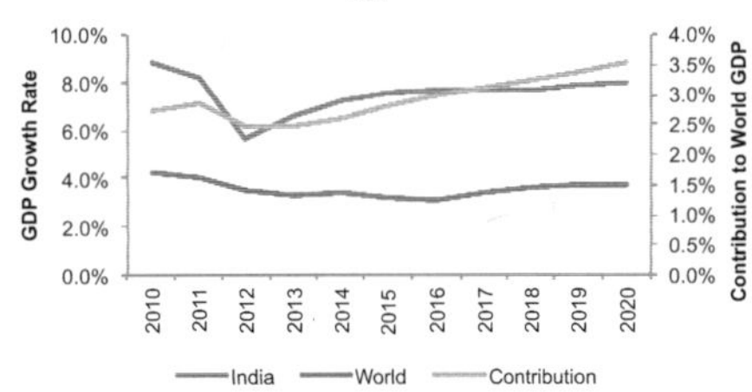

Global Corporations Investing Large Sums in Creating Manufacturing Capabilities in India

India – FDI in US$bn from 2012 to 2018 (Expected)

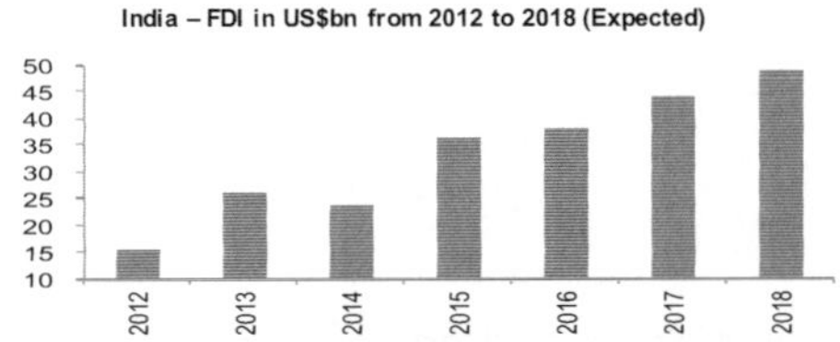

Falling Interest Rates Will Lead to Higher CAPEX and Infrastructure Developments

Reserve Bank of India – Repo Rate

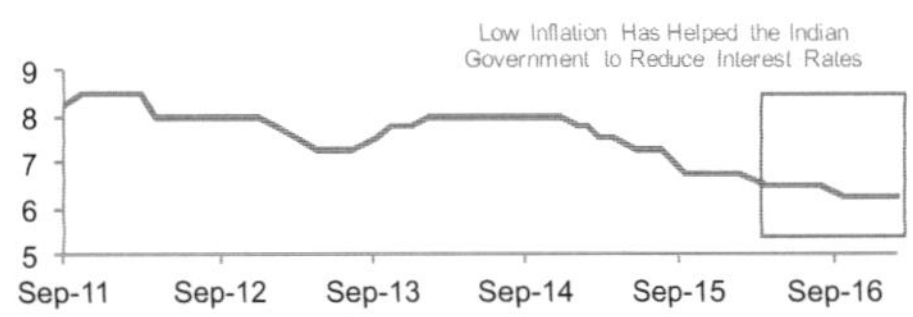

Rising GDP, FDI coupled with start of lower interest regime rate augurs well for the Indian economy as it embarks on becoming the next global super-power

Source: Bloomberg, CIA Fact Book, World Bank, IMF, Ministry of Finance - India

Significant Challenges Being Faced by the Indian Economy

JG CHEMICALS

"India will be a super power....we will invest in all the key elements to increase growth and attract FDI. India will be the manufacturing hub of the world" *Narendra Modi, Indian Prime Minister, 2016*

Issue	Description	Government Action
Lack of Infrastructure	▪ The lack of basic infrastructure like roads, ports, access to water etc have hampered India's growth; red-tapism delayed several key projects	Creation of Infra Fund Dedicated for Key Sectors
Access to Capital and Financing	▪ The Indian banking system suffered immensely during the last 4 years due to the impact on key sectors like infrastrcuture, steel and power	Reduction of Interest Rates and Recapitalisation of Banks
Complex Tax Structure	▪ Multiple state and central taxes, duplication of taxes and high rates of corporate income tax have led to a strong parallel "black" economy	Introduction of Goods and Services Tax in July 2017
Governance	▪ Corruption across all levels of government led to the creation of a negative impact in front of global corporations	Focus on Clean Governance and Single Window System for Clearances

Like most emerging markets, the Indian economy does face a few challenges; however, the current government has initiated adequate steps to address each of them

Source: Bloomberg, Business India, The Telegraph, KPMG

JG CHEMICALS

Table of Contents

Agenda

JG CHEMICALS

1. **Technology**
2. Production
3. Consumption
4. International Trade
5. Raw Material Scenario
6. Regulatory Environment
7. Future Outlook

1. Technology: French Process Continues to Dominate

JG CHEMICALS

Key Characteristics

- **French Process:** In India close to 78% of all zinc oxide production is undertaken through the French process technology
- **Calcination Process:** This process accounts for roughly 16% of the total zinc oxide production in India; only two producers are manufacturing through this technology. Capacity growth is limited due to RM constraints
- **Wet Process:** This technology accounts for roughly 6% of the total production; mainly used to manufacture highly specialized grades of Active zinc oxide and feed grade zinc oxide (exports)
- **American Process:** This technology is prevalent only in a very crude form and hence its contribution is negligible. This is a very low quality ZnO; more like a grey oxide

Technology Wise Production

Technology Wise Segmentation of Total Zinc Oxide Production in India for 2016

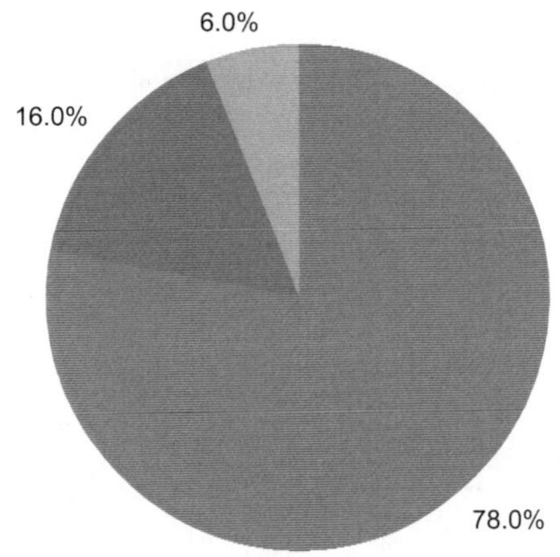

Majority of the zinc oxide production in India continues to be through the French Process; minor increase has been witnessed in the share of the Wet Process in the last 2-3 years

Source: JGC Research

1. Technology: Technological Up-gradation Required

JG CHEMICALS

- **Lack of Differentiation.** Fairly standard technology being adopted by most producers; very little differentiation in plant design and technology even with the French Process.
- **Low Level of Investments.** Most producers are small and hence unable to undertake large scale capital investments in R&D and technological up-gradation; low capacities further prevent investment in R&D activities
- **Need for Global Knowledge Transfer.** Indian producers need to reach-out to global leaders to engage in technology transfer to upgrade existing manufacturing systems and processes
- **Absence of Partnership Between Academia and Industry.** Lack of partnership between academia and industry has led to very little research on new technologies and grades in zinc oxide

Indian producers need to focus on increasing their existing capabilities by engaging in global knowledge transfer and focus on R&D; consolidation will help channel funds in these areas

Source: JGC Research

Agenda

JG CHEMICALS

1. Technology
2. **Production**
3. Consumption
4. International Trade
5. Raw Material Scenario
6. Regulatory Environment
7. Future Outlook

2. Production: Total Production Capacity in India

JG CHEMICALS

Key Statistics

Total Manufacturers:	50-60 (Average 55)
Total Installed Capacity:	130,000MT
Current Operational Capacity:	110,000MT
Surplus Capacity:	20,000MT

Capacity Utilization

Capacity Utilization and Surplus Capacity from 2006 to 2012

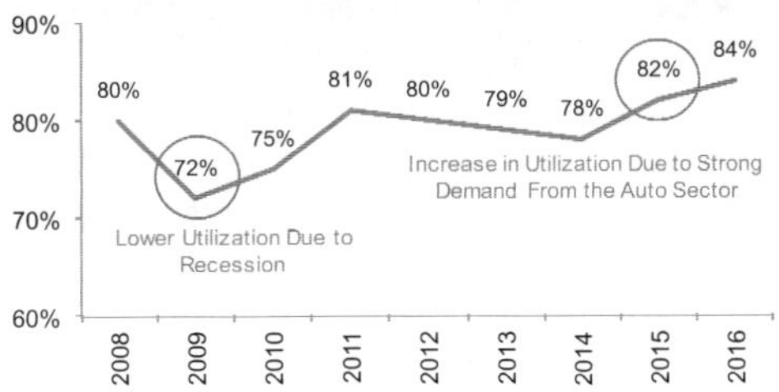

Evolution of Zinc Oxide Production

Total Zinc Oxide Production in India from 2008 to 2016

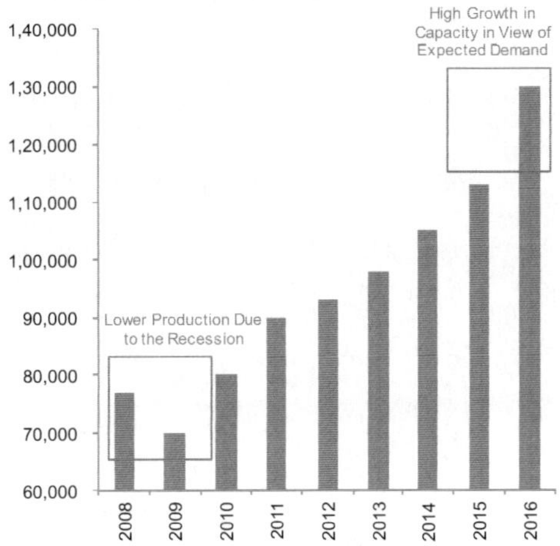

Significant capacity additions have taken place during the last 3 years in view of the expected increase in demand; further capacity additions will lead be detrimental for industry

Source: JGC Research Note: All production related figures are annual

A BDJ GROUP ENTERPRISE

2. Production: Highly Fragmented Landscape

JG CHEMICALS

Key Features

- **Highly Fragmented Landscape:** Average production per manufacturer is ~167MT per month and excluding top 3 players is only 103 MT per month
- **Pan India Presence:** Multiple manufacturing facilities are located in every region of India
- **Need for Consolidation:** There is an urgent need for consolidation in the market
- **Irregular Production Cycle:** Most small manufacturers work for an average of 6 months in a year

Market Share

Domestic Market Share for Top 3 Manufacturers Vs Remaining Manufacturers for 2016

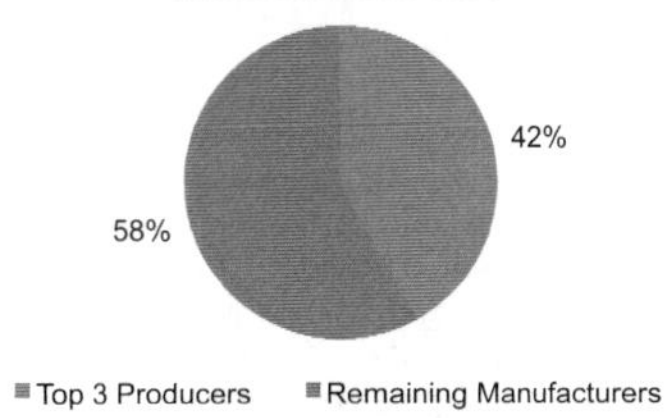

Average Production (All Manufacturers)

Sales Tonnage per Manufacturer (Including all Manufacturers)

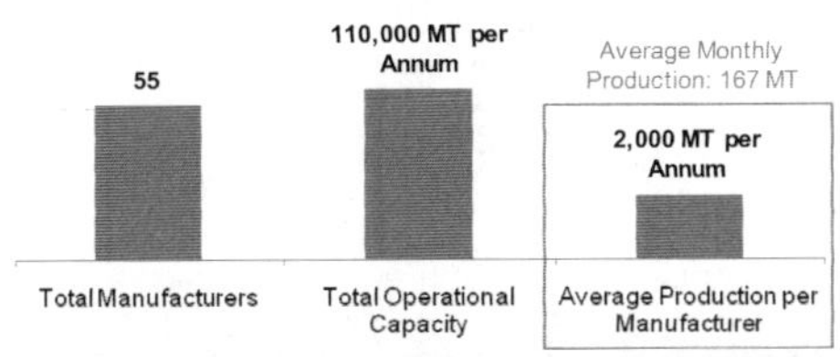

Average Production (Excluding Top 3)

Sales Tonnage per Manufacturer (Excluding Top 3 Manufacturers)

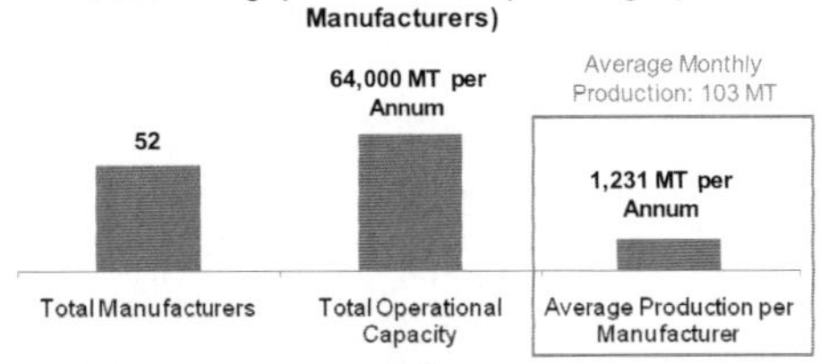

Highly fragmented industry with average production of around 100MT per month apart from the top 3 producers; most of the smaller producers are seasonal producers and not managed with a professional approach

Source: JGC Research

A BDJ GROUP ENTERPRISE

Agenda

1. Technology
2. Production
3. **Consumption**
4. International Trade
5. Raw Material Scenario
6. Regulatory Environment
7. Future Outlook

3. Consumption: Across a Wide Range of Industries

Key Statistics

Total Market Size:	**1,00,000MT**
Domestic Share :	**97,600MT**
Import Share:	**2,400MT**

Key Characteristics

- **Highly Dependent on the Rubber Sector:** Over 70% of the total consumption is used in the rubber and tyre industry;
- **Increasing Consumption in Ceramics:** From roughly 15% in 2011 to the current ~20% of total consumption; trend expected to continue due to the demand for housing
- **New Usage of Active and SHG Grade:** Certain industries have started using active zinc oxide, although in a very small manner. SHG zinc oxide demand is increasing in the specialty chemicals and pharma sector

Global Consumption Pattern

Global Zinc Oxide Consumption - 2016

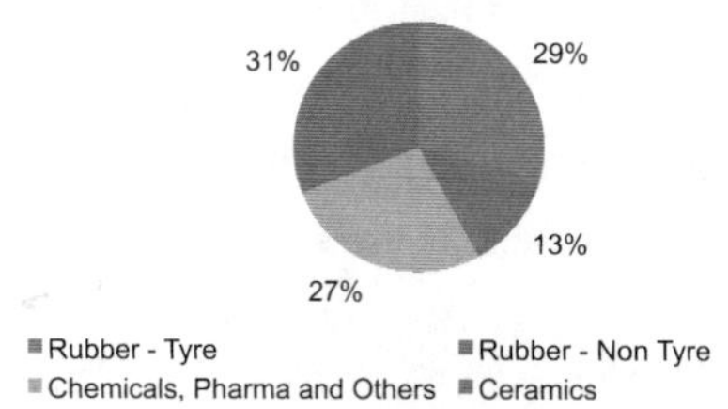

Indian Consumption Pattern

Indian Zinc Oxide Consumption - 2016

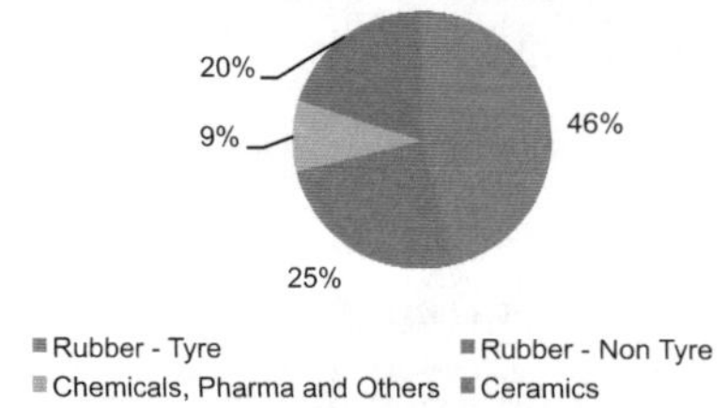

Majority of the zinc oxide consumption is centred around the rubber (tyre and non tyre) and ceramics; other areas are still very small; growth is visible across all sectors

Source: JGC Research, AIRIA, ILZDA, IZA

3. Consumption: High Growth Potential

JG CHEMICALS

Sectors		Growth Drivers
Rubber Products		▪ Low Per Capita Consumption: Per capita rubber consumption is only 1.16 kg which is significantly lower than the 14-16 kg in US and Europe ▪ Export Focus: India's rubber product manufacturers are focussing on producing value added products for export markets in the West
Tyres		▪ India's Development as an Automobile Hub: Global auto majors as well as domestic companies have invested over US$8bn over the last few years; further expansions underway ▪ Rise in Tyre Production: India's tyre production has risen from 60mn units in 2004-05 to 150mn units in 2015-16 implying a CAGR of c8%
Real Estate (Ceramics and Paints)		▪ Growth of the Indian Real Estate Industry: The real estate industry is expected to register strong growth in the future due to the governments' focus on "housing for all" by 2022 target ▪ Demand from Rural India: Rising middle and lower income segments of the population has led to an additional demand from rural areas
Agriculture / Fertilizers		▪ IZA Initiative on Use of Zinc Oxide in Fertilizers: IZA has been working closely with the Indian government to promote the use of zinc oxide in agriculture / fertilizers. Unlike in the West, this is a large and currently untapped segment which has a strong potential for growth

Zinc oxide consumption in few industries is expected to witness high growth due to the buoyant future growth potential of the Indian economy in general

Source: Bloomberg, Press Trust of India, Kotak

3. Consumption: Key Challenges for Manufacturers

JG CHEMICALS

- **High Bargaining Power of Customers.** Large number of producers chasing few large customers results in higher bargaining power and options for customers
- **Price Sensitive Market.** Due to the commodity nature of the product, price is often the only criteria for vendor selection
- **Reduction of PHR.** Due to rising costs, tire and other rubber companies are trying to cut down on the use of ZnO; evident as ZnO demand has not increased as much as tire production has
- **Long Vendor Approval Process.** Vendor registration and approval with large tyre companies is a very complex and time consuming (3-4 years) process
- **Absence of Long Term Partnership Principles.** Lack of long term commitment from key buyers in terms of pricing formulae and quantity

Agenda

1. Technology
2. Production
3. Consumption
4. **International Trade**
5. Raw Material Scenario
6. Regulatory Environment
7. Future Outlook

4. International Trade: Low Level of Exports and Imports

Key Characteristics

- **Imposition of Anti Dumping Duty:** Imports of zinc oxide significantly dropped after the imposition of anti-dumping duties on Chinese producers by the Indian Govt.
- **Higher Grades are Usually Imported:** Higher quality grades which are used in more advanced and specialized applications are usually imported by Indian consumers
- **Export Grade:** Most of the Export tonnage is of the feed grade of zinc oxide
- **Volatility in LME & Exchange Rate:** High volatility in zinc prices in the LME coupled with exchange rate fluctuation is a deterrent to export

Key Statistics (for 2016)

Total Exports:	**12,400MT per annum**
Total Imports:	**2,400MT per annum**

Key Export Destinations

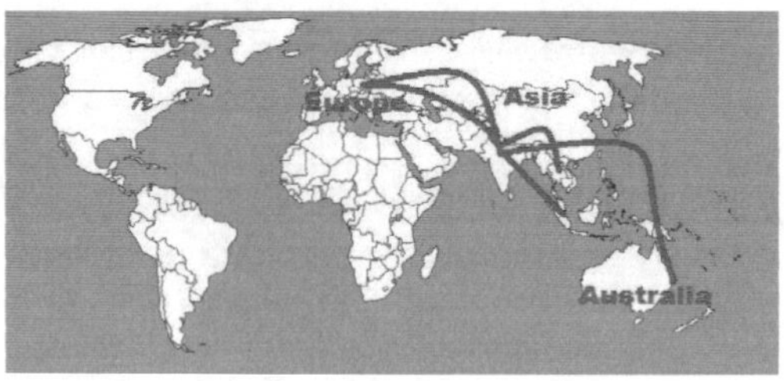

1. **Asia**
2. **Australia**
3. **Europe**

Source: JGC Research, Government of India, KPMG, Bloomberg, ILZDA

Agenda

JG CHEMICALS

1. Technology
2. Production
3. Consumption
4. International Trade
5. **Raw Material Scenario**
6. Regulatory Environment
7. Future Outlook

5. Raw Material Scenario: Key Features

JG CHEMICALS

Key Features

- **Only 1 Producer:** Hindustan Zinc (HZL) is the only producer of zinc metal in the country with a current production of about 800,000 MTPA
- **Increasing Imports of Zinc Metal:** Share of import has increased due to the issues with HZL's mines in the last couple of years
- **Slowdown in the Galvanizing Industry.** Falling infrastructure spending due to high interest rates has caused a slowdown in the output; this has led to lower availability of zinc scrap; however, this trend is expected to reverse in the coming years
- **Dependence on Imports for Zinc Scrap.** Large producers like JGC have to be dependent on imports to maintain continuous production. JGC is today the largest importer of zinc scrap in India
- **Lack of Co-relation Between LME and Domestic Secondary Zinc.** Prices of domestic secondary zinc dross does not move in tandem with the LME prices and hence it is difficult to hedge positions in commodity exchanges

Consumption Pattern in Zinc Oxide

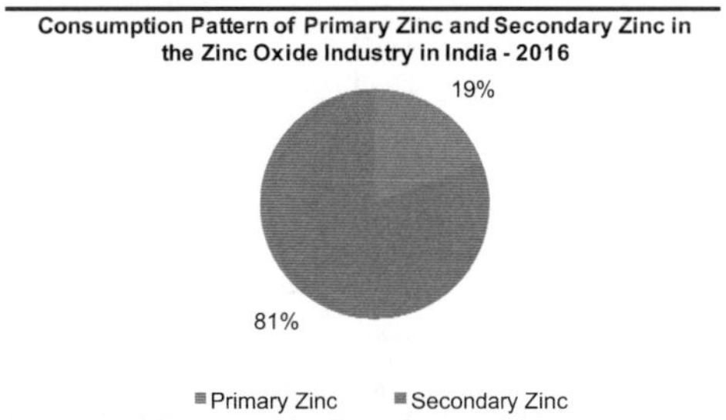

Share of Domestic Producers Vs. Imports

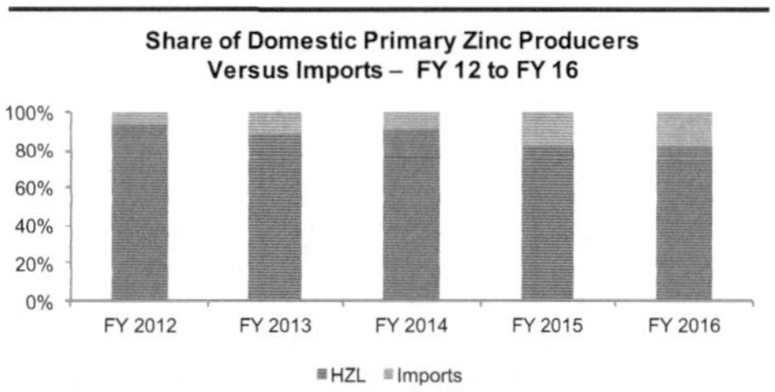

Secondary zinc is the major raw material for zinc oxide production in India; with zinc metal prices remaining volatile, the availability and pricing of secondary zinc has become difficult

Source: JGC Research, Bloomberg, ILZDA, IZA

Agenda

JG CHEMICALS

1. Technology
2. Production
3. Consumption
4. International Trade
5. Raw Material Scenario
6. **Regulatory Environment**
7. Future Outlook

7. Regulations: Complex and Uncertain Environment

JG CHEMICALS

Market Landscape

- **Frequent Changes and Lack of Clarity in Domestic Policies and Tariffs.** Lack of continuity in domestic regulations have disrupted the growth of the industry due to project cancellations and reduced domestic demand
- **Stricter Environmental Regulations.** Stringent and complex norms are being introduced with respect to pollution control and environmental protection
- **Delay in Getting Environmental Clearances.** Despite large scale capital investments and environment protection equipments, clearances and permits for new units take years
- **Lack of Single Window Clearance.** Multiple clearances are required from both state and central Government authorities resulting in a significant wastage of time, man-power and resources
- **Poor Governance Across the Unorganized Segment.** A large unorganized segment continues to grow and develop despite flouting almost all environmental and legal norms; poor enforcement and corruption have prevented curbs on their activities

Regulatory framework is fairly complex; a European producer who entered the Indian market had suffered due to large tax liabilities and inability to adapt to the Indian market requirements

Source: JGC Research, Press Trust of India

Agenda

JG CHEMICALS

1. Technology
2. Production
3. Consumption
4. International Trade
5. Raw Material Scenario
6. Regulatory Environment
7. **Future Outlook**

7. Future Outlook: Growth Expected to Continue

JG CHEMICALS

- ❑ **Increase in Demand for Zinc Oxide.**
  - Automotive sector in India to grow due to announcement of new projects by MNCs
  - The Indian economy expected to surge by clocking GDP growth rate of over 7% in the next 5 years
  - The Government's "make in India" campaign will further drive the growth of manufacturing in India
- ❑ **Further Capacity Additions will be Detrimental – Spare Capacity Already Exists.**

Prime Minister, Narendra Modi launched "Make in India" on 25 September 2014 in a function at the Vigyan Bhavan..... The initiative hopes to attract capital and technological investment in India.

Source: www.makeinindia.com

The Indian zinc oxide industry is expected to grow on the back of increased tyre demand; however, capacity additions will be harmful for the industry in the current scenario

JG CHEMICALS

Thank You

Science Advocacy:
Regulatory Developments and Actions in North America

Eric Van Genderen | International Zinc Association
1 March 2017 | Rancho Mirage, CA

International Zinc Association

Regulatory Pressures

- US States
  - CA: "Zinc in Tires" and Hazardous Waste issues
  - WA: Restricted use in new developments
- US National
  - Water quality updates
  - Toxic Substances Control Act (TSCA)
  - USEPA nano regulations
- Canada
  - Environment and Climate Change Canada (ECCC)
  - Chemical Management Plan (CMP)

International Zinc Association

"One of These Things is Not Like the Others…"
- Washington, 2015 -

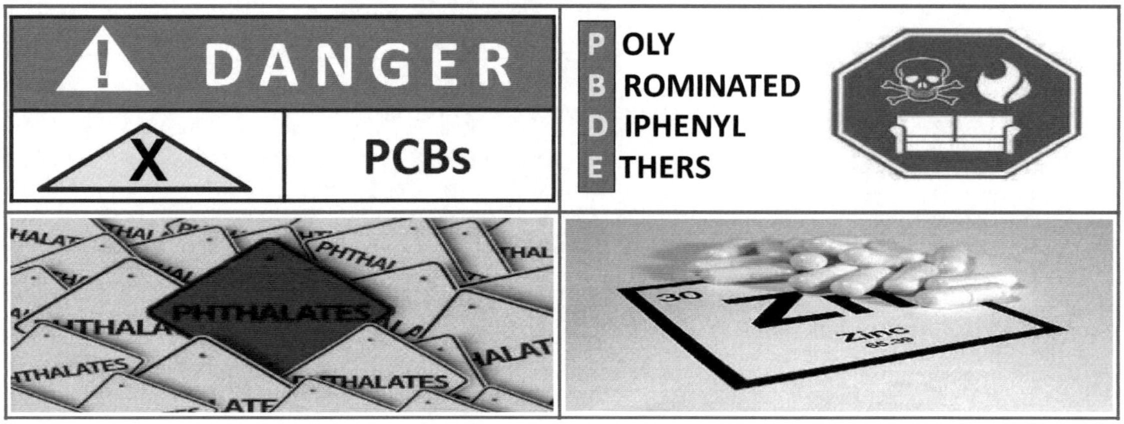

International Zinc Association

"One of These Things is Not Like the Others…"
- Washington, 2015 -

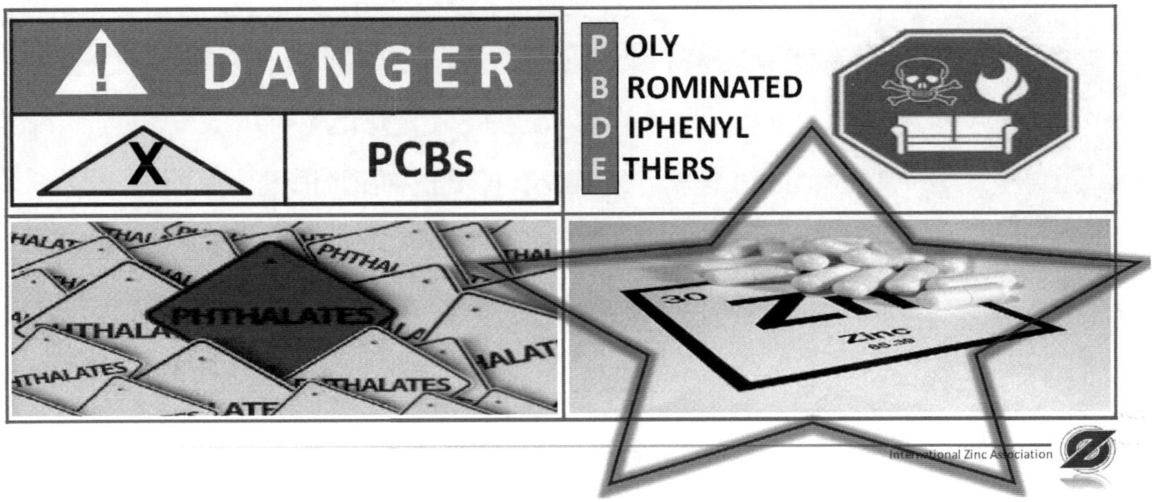

"One of These Things is Not Like the Others…"
- California, 2016 -

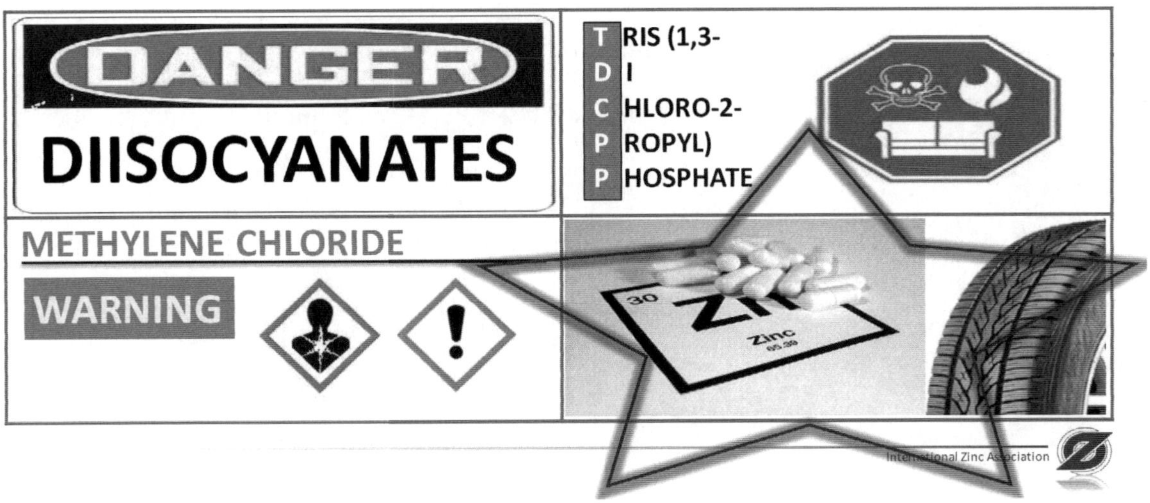

The Left Coast

- California (CA), Oregon (OR), Washington (WA)
- Progressive environmental agencies
- Policy migration into other states:
  - ➢CA: Priority Product regulations
  - ➢OR/WA: Protect trout and salmon at all costs

CA Attempts to Restrict Zn in Tires

- Municipality facing penalties for "impairment"
  - ➢ Local Senator drafted Bill to eliminate Zn in tires
  - ➢ Zinc Coalition mobilized (Bill defeated in <30 days)
- IZA working with Water Boards to avoid DTSC petition for zinc/tires as Priority Product

California Environmental Protection Agency | Department of Toxic Substances Control | California Water Boards

International Zinc Association

California DTSC

Department of Toxic Substances Control

- Challenging recycling criteria
  - ➢...must go to primary smelter
  - ➢...must not be used in land application
- DTSC hesitant to recognize practical aspects of zinc recycling without change to statutes
  - ➢Recycling process, fertilizers, etc.

International Zinc Association

Washington Restrictions

- Roofing concerns disproven with Ecology's own research
- Rigorous BMP requirements for new development
  - ➢Galvanized bridges and building facade

International Zinc Association

North America

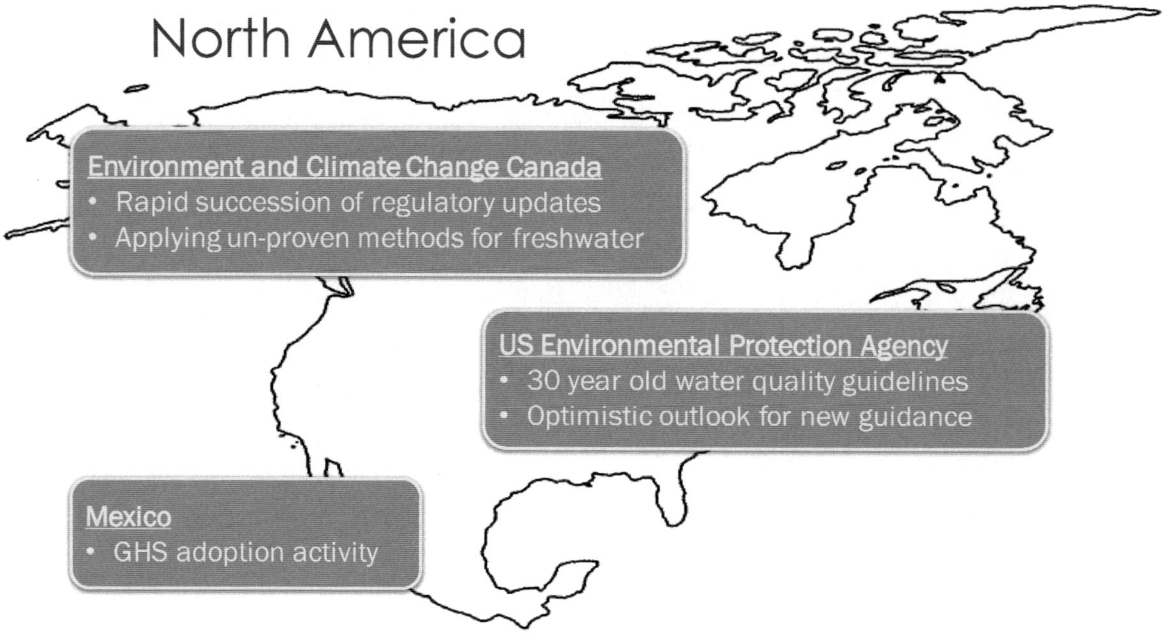

Biotic Ligand Model

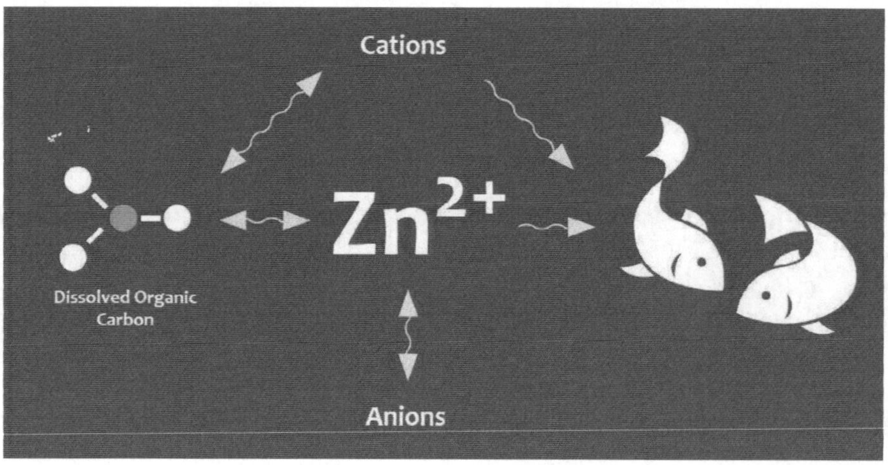

International Zinc Association

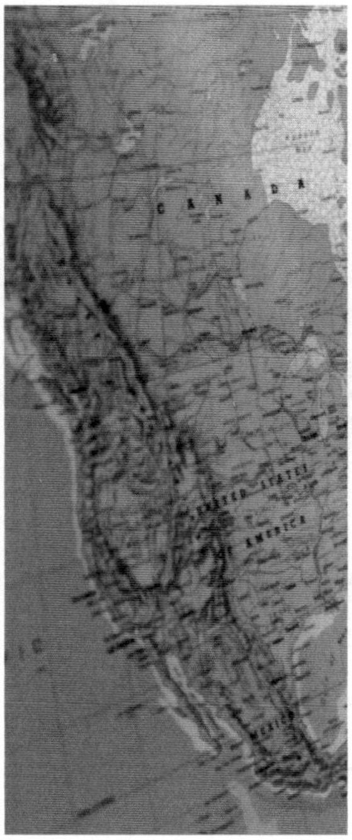

North American Activities - USA -

- Advising USEPA on Zn-BLM guidance
  - ➢ Validation studies with Region 5 (Great Lakes)
  - ➢ CRADA between USEPA and associations to develop guidance for metals
- Stormwater monitoring study with galvanizers
  - ➢ Working with AGA to update USEPA stormwater benchmarks

International Zinc Association

Toxic Substance Control Act (TSCA)

Companies that manufacture or import a chemical substance in 2012-2015 (inclusive) that:

- Is subject to the TSCA

AND • Is listed in the TSCA Inventory

AND • Is not otherwise exempt from CDR reporting

AND • Meets or exceeds the CDR reporting threshold

➢Threshold depends on substance and company size

International Zinc Association

ZnO Nano

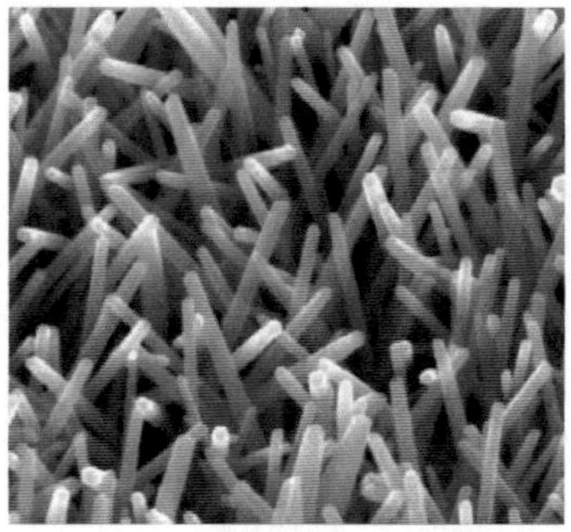

US Federal Register (80) April 6, 2015

EPA is proposing to exclude from the requirements of this rule nanoclays, zinc oxide and chemical substances manufactured at the nanoscale as part of a film on a surface.

US Federal Register (82) January 12, 2017

EPA did not include the proposed exemptions for nanoclays and zinc oxide in the final rule.

International Zinc Association

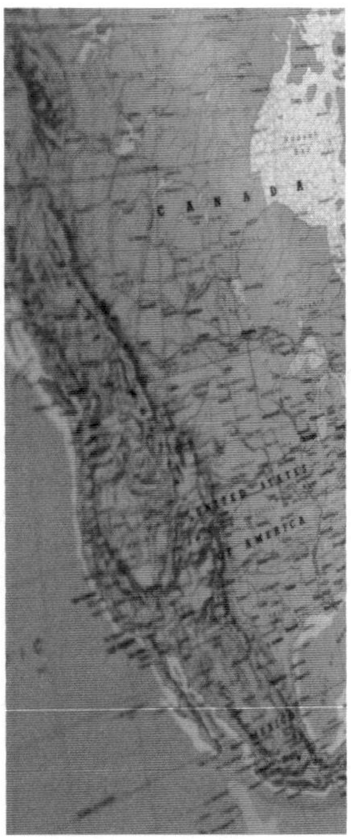

North American Activities - Canada -

- Final Water Quality Guideline expected in 2017
  - Draft contained significant errors, request for complete withdrawal denied
- Draft Soil Quality Guideline released in 2016
  - SQG would increase 20% (430 mg/kg)
- Preparing for "Zinc Group" actions under CMP
  - Assessment being prepared in 2017
  - EU REACH Dossier will be important contribution

International Zinc Association

Take Away

- IZA is a resource for regulatory action:
  - ➢Develop platforms for scientific opposition
  - ➢Conduit for delivering stakeholder messages to the political fray
- Harmonized chemical management preserves market access
- Proliferation of regulatory updates over next few years
- Help us, help you!

International Zinc Association

Thank You!

Eric Van Genderen, Ph.D.
evangenderen@zinc.org
international zinc association

International Zinc Association

Zinc Nutrient Initiative:
Update on Impacts &
Activities
Zinc Nutrient
Initiative
Dr. Andrew Green
Zinc Oxide Industry Conference
Palm Springs, CA, USA
March 1, 2017

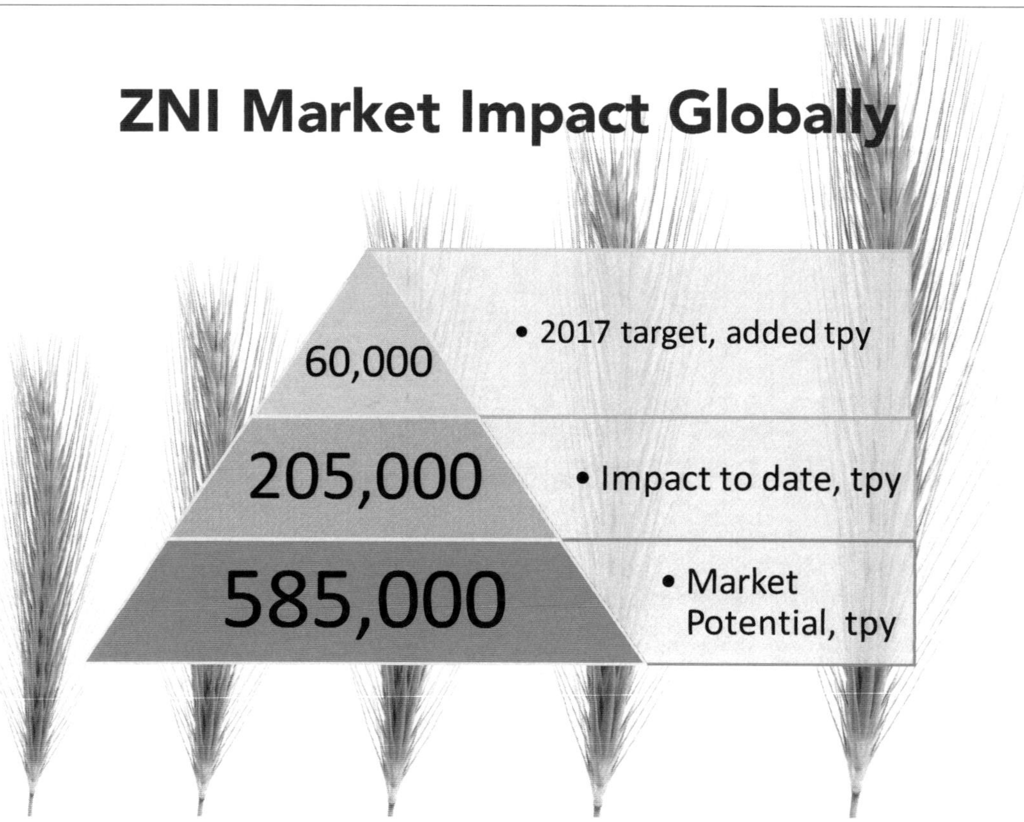
ZNI Market Impact Globally
60,000
• 2017 target, added tpy
205,000
• Impact to date, tpy
585,000
• Market Potential, tpy

ZNI Activities

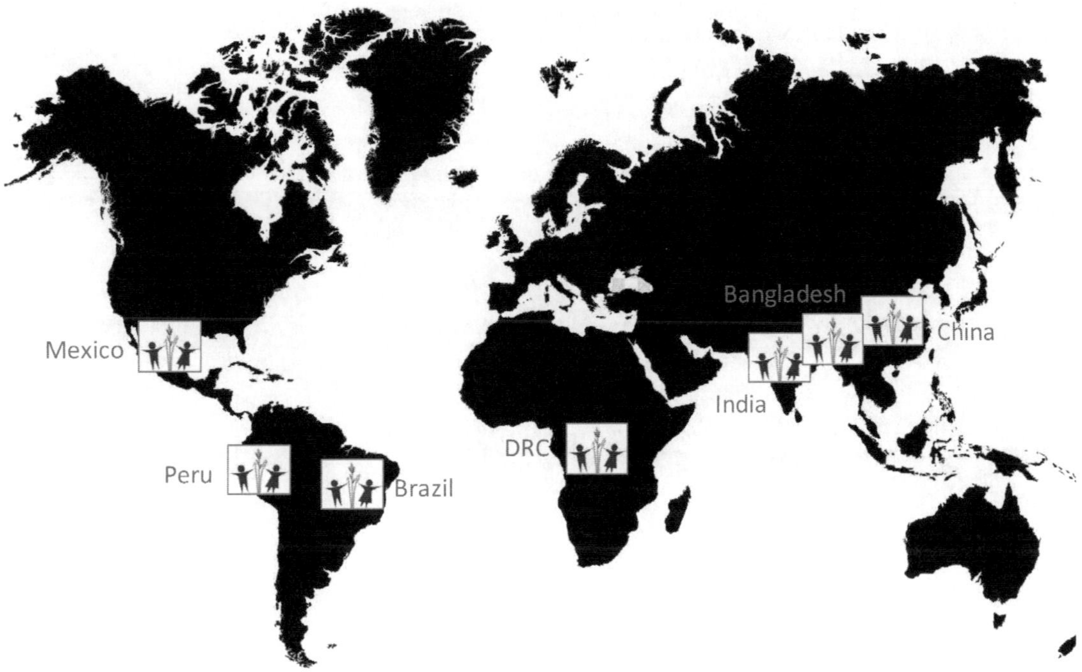

India
Impacts in 2016

- IFFCO - $ZnSO_4$ mono increased to 45,000 tons (from 30,000 tons)
- ANYA (KRIBHCO) – New 16,500 tons $ZnSO_4$ hepta + mono
- 20% increase in Zn-fertilizer use (≈ 12,000 tons Zn-metal)
- 5 new grades of Customized Fertilizers containing Zinc
- Inclusion of new Zn-fertilizers in Fertilizer Control Ordinance

International Zinc Association

Zn-fertilizer recommendation increased

S.No	Crop	State	Zn-recommendation
1	Rice	Punjab	62 kg $ZnSO_4$/ha
2	Rice	Andhra Pradesh	50 kg $ZnSO_4$/ha
3	Rice	Madhya Pradesh	50 kg $ZnSO_4$/ha
4	Wheat	Madhya Pradesh	50 kg $ZnSO_4$/ha
5	Wheat	Punjab	62 kg $ZnSO_4$/ha
6	Soybean	Madhya Pradesh	50 kg $ZnSO_4$/ha
7	Pigeon pea	Madhya Pradesh	50 kg $ZnSO_4$/ha
8	Ground nut	Punjab	62 kg $ZnSO_4$/ha
9	Ground nut	Maharashtra	50 kg $ZnSO_4$/ha
10	Ground nut	Andhra Pradesh	50 kg $ZnSO_4$/ha
11	Ground nut	Telengana	50 kg $ZnSO_4$/ha
12	Ground nut	Rajasthan	30 kg $ZnSO_4$/ha
13	Sugarcane	Tamil Nadu	37.5 kg $ZnSO_4$/ha
14	Cotton	Tamil Nadu	50 kg $ZnSO_4$/ha
15	Tomato	Tamil Nadu	50 kg $ZnSO_4$/ha

(General recommendation: 25 kg $ZnSO_4$/ha)

International Zinc Association

$ZnSO_4$ consumption trend

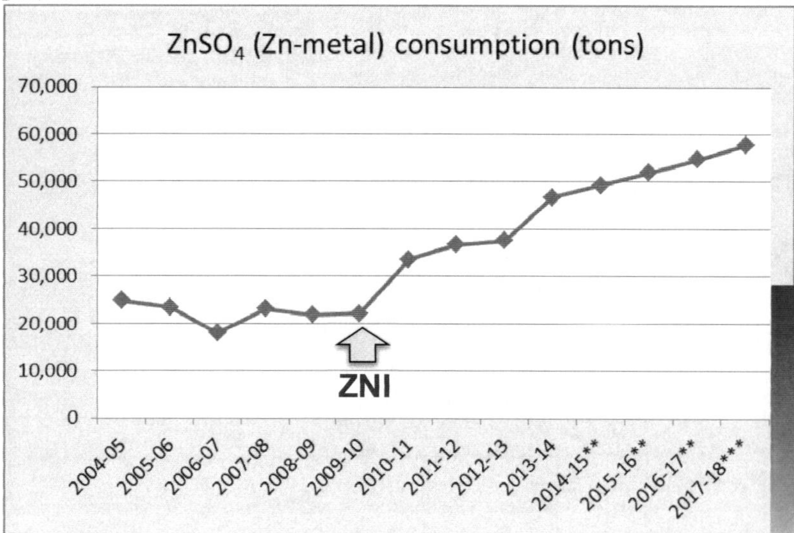

**ZNI estimate
***Projection

Source: FAI (2016)

International Zinc Association

INDIA: Next Steps

- Policy intervention on Zincated Urea
- Change in critical deficiency level
- Updated deficiency maps/recommendations
- Inclusion of ZnO powder in FCO
- Zn fertilizer app

International Zinc Association

Bangladesh Project

- **IZA-BARI Project on "Promotion of Zinc Fertilizer Use in Bangladesh for Food and Nutrition Security" initiated in May 2016**
- **MOU signed between IZA and BARI in February 2017**
- **Market increase is ~ 5000 tons Zn-metal in 2016**

International Zinc Association

CHINA: Key Developments

- 30% increase in the zinc fertilizer market in China in the last year
- 40% increase of Zn-containing compound fertilizer market
- 50% increase of ZnSO4 + ZnO consumption as fertilizer last year
- Increasing total capacity of ZnSO4 to 650 KMT; total production to 580 KMT.

International Zinc Association

Zn Metal Consumption in China

(1000 tons)

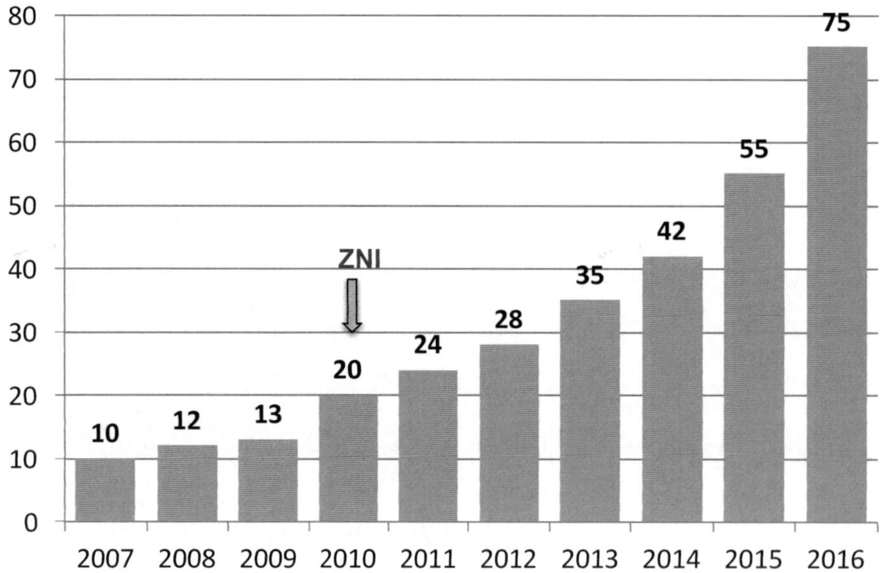

International Zinc Association

Market Driving Food & Nutrition Security

- China program has increased Zn fertilizer market by 75,000 tons
- This has resulted in:
  - 24 million tons of Zn-enriched food
  - Improved nutrition for more than 120 million people

CHINA: Next Steps

- More Zn-containing fertilizers will come to market as project with MoA progresses
- Zinc will be added to additional provincial fertilizer formulations and formula fertilizer production
- Zinc fertilizer consumption to increase by another 20,000 tpy in 2017

International Zinc Association

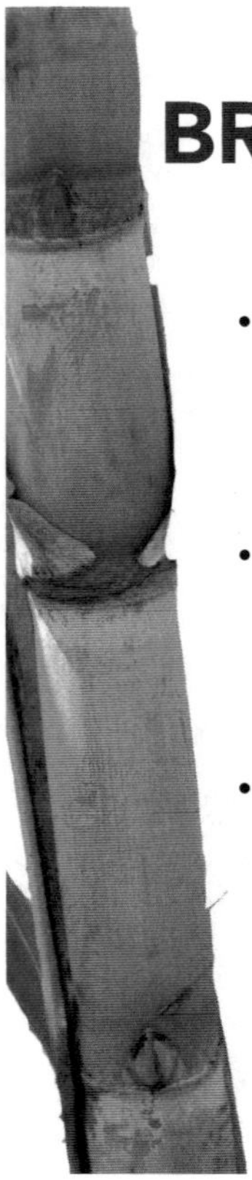

BRAZIL: Developments

- New Zn deficiency map created for Brazilian soils
- It is estimated that up to 52% of soils are deficient
- New "seal of approval" program being developed

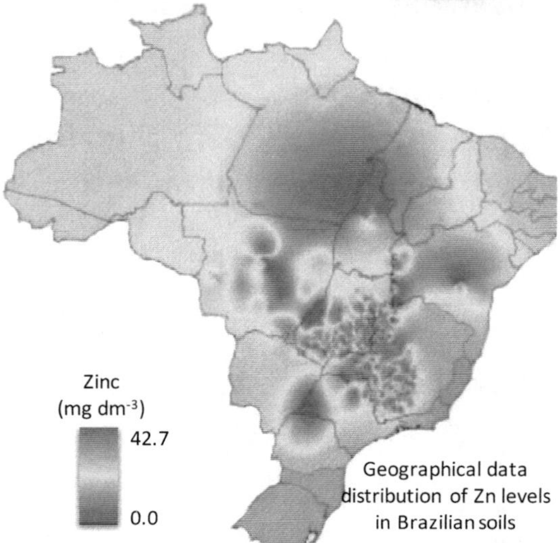

International Zinc Association

MEXICO: Developments

- Project with Instituto Nacional de Investigaciones Forestales,Agrícolas y Pecuarias (INIFAP) with goal of developing goverment recommendation for zinc fertilizer
- Final report will be completed in 1st half of 2017
- Other Activities:
  - Supporting major fertilizer conference in Mexico
  - Series of promotional workshops being held
  - Reaching out to government to discuss policy changes

International Zinc Association

Democratic Republic of the Congo

- Zinc crops program underway with Lundin Foundation
- Focused on improving food & nutrition security with local farmers
- 3 year, $500,000 program

Zinc Association

DRC: New Project

- New zinc crops pilot project with MMG
- Long-term project to be considered if pilot is successful
- 5 crop trials underway near Kinsevere mine

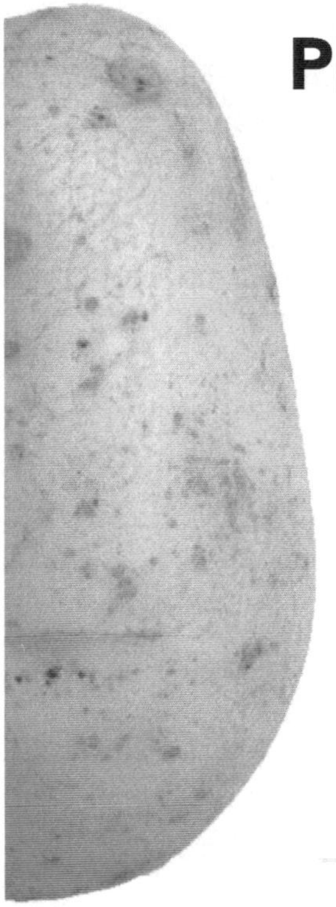

PERU: New Project

- Working with the International Potato Institute and Universidad Nacional Agraria La Molina
- 4 new trials underway
- Goal is to improve potato production and nutrition

International Zinc Association

World Bank Initiative

- Meeting held at World Bank HQ in December, 2016
- IZA invited back for high-level meeting in April/June
- Paper planned at World Bank Africa meeting as well
- Goal: Incorporation of zinc fertilizer in World Bank global agriculture programs

THE WORLD BANK
IBRD • IDA

:iation

Current ZNI Activities

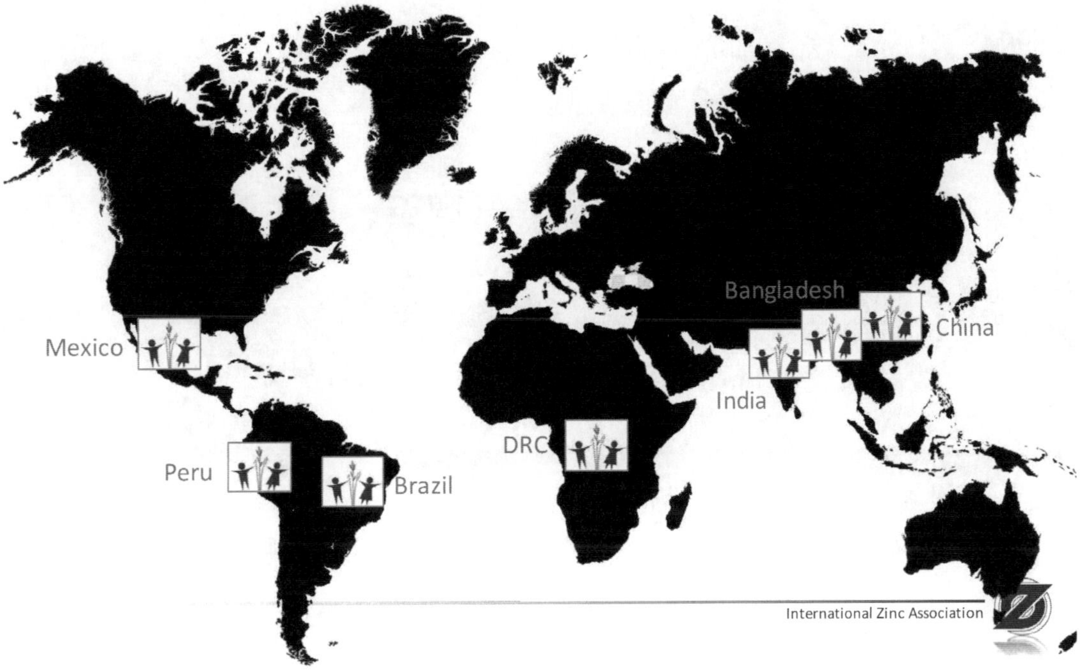

International Zinc Association

World Bank Activities

> 12,000 projects in 173 countries

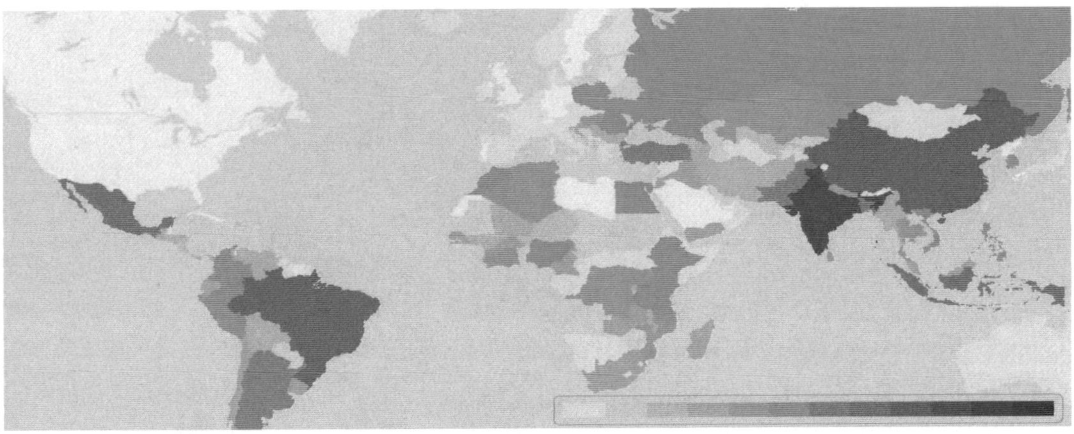

International Zinc Association

32 ZNI Affiliate Members

CONCLUSION

- ZNI program has achieved significant market impacts to date
- Ongoing activities expected to increase market impact even further
- ZNI also helping with global food security, nutritional security, and human health

Zinc Nutrient Initiative

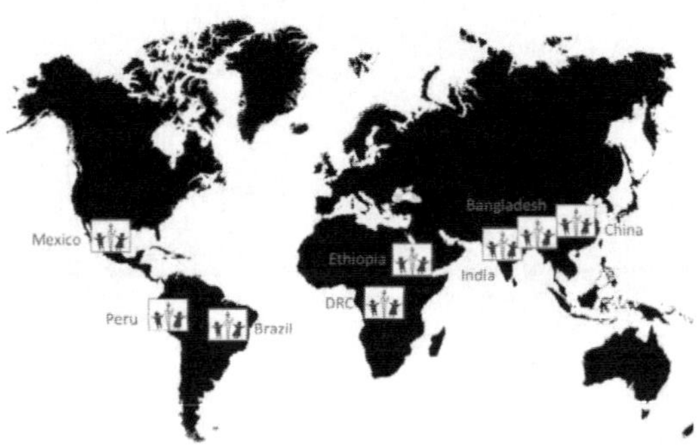

International Zinc Association

Thank You
Zinc Nutrient
Initiative
www.crops.zinc.org

Zinc Market Review: India

Frank E. Goodwin Sc.D.
International Zinc Association

2017 International Zinc Conference & Zinc Oxide Industry Conference

International Zinc Association

India Market Statistics – Primary Zinc

Primary Zinc, '000t	2011-12	2012-13	2013-14	2014-15	2015-16	2016-17, E	2017-18, E
Domestic Supply	465	491	567	488	525	464	560
Imported Supply	64	83	53	129	128	202	125
Total Primary Consumption	529	574	620	617	653	666	685
Domestic Share: Primary Zinc	88%	86%	91%	79%	80%	70%	82%
Export: Primary Zinc	306	204	182	248	135		

Source: HZL

International Zinc Association

India pattern of zinc consumption

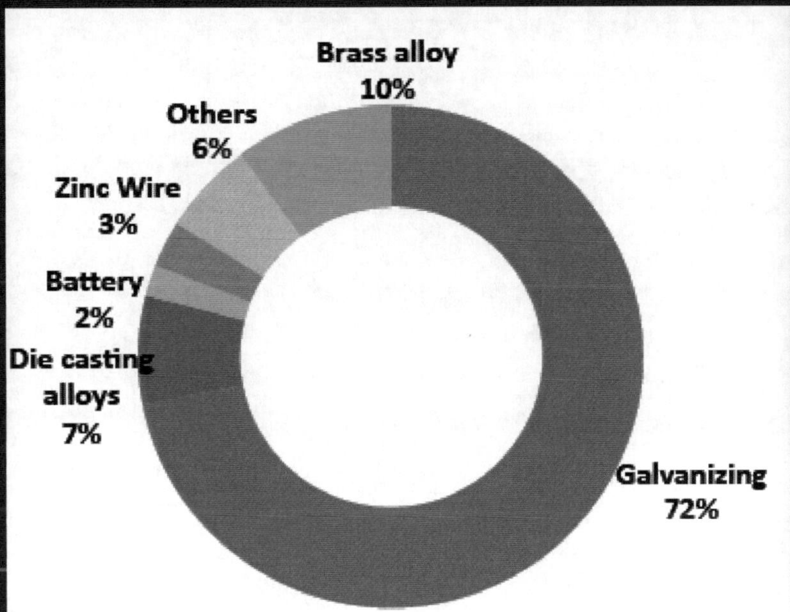

Source:ILZSG

India Market Statistics – Zn Diecasting Alloy

Zn Diecasting Alloy, tonnes	2011-12	2012-13	2013-14	2014-15	2015-16	2016-17, E	2017-18, E
Domestic Supply	3	7	9	4	0	1	15
Import	6	8	18	28	25	26	18
Total	9	15	27	32	25	27	33
Domestic Share	33%	47%	33%	13%	0%	4%	45%

International Zinc Association

Source: HZL

India Market Statistics – Scrap/Other

Import Type, tonnes	2011-12	2012-13	2013-14	2014-15	2015-16	2016-17, E	2017-18, E
Import : DieCast Scrap	13	20	20	23	18	19	19
Import : Scrap/ Dross	30	35	42	52	35	45	43
Import : Zn Dust	3	3	3	4	4	4	4
Import : Others	4	4	2	2	2	2	2
Import : Ash	24	26	34	31	23	31	30
Import : Brass	45	62	39	48	50	54	50
Import : Oxide	5	3	4	5	4	4	4
Total Scrap/Other Imports	124	153	144	165	136	159	152

International Zinc Association

Source: HZL

India Market Statistics – Downstream Zn Units

Import Type, tonnes	2011-12	2012-13	2013-14	2014-15	2015-16	2016-17, E	2017-18, E
Import : Galvanized sheet	16	20	18	22	38	21	22
Import : Diecast product	22	18	18	22	22	25	23
Total Downstream Zn Imports	38	38	36	44	60	46	45

International Zinc Association

Source: HZL

India Domestic Market Size, tonnes

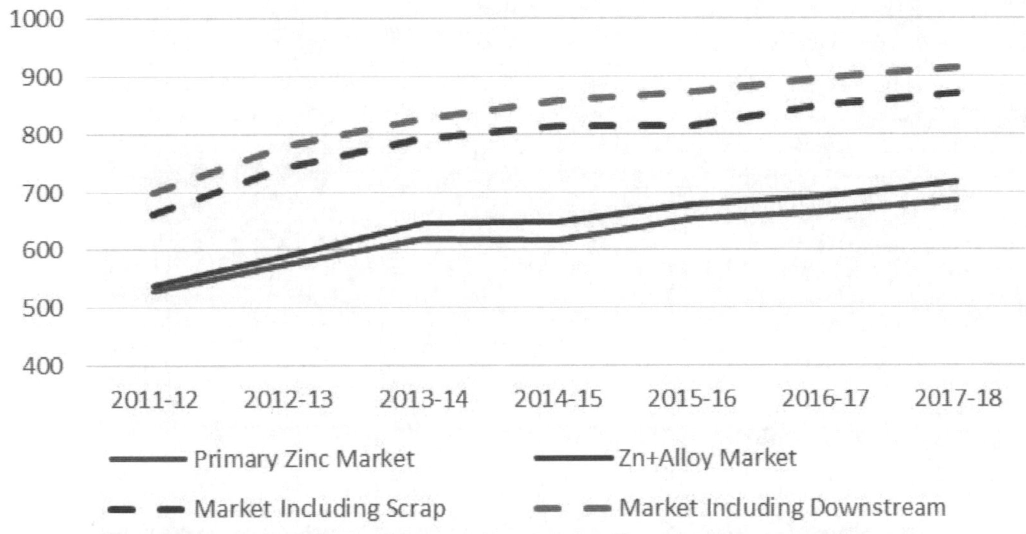

International Zinc Association

Source: HZL

India Market Statistics – Zn and Galvanized Sheet

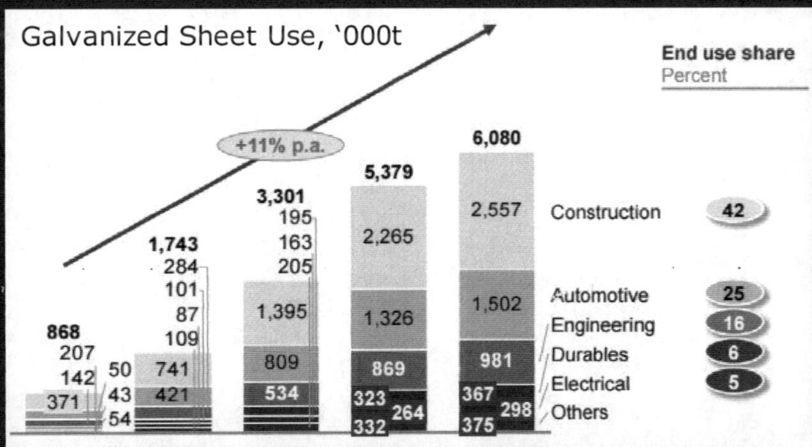

Zinc Use '000T	11	22	41	67	75	86	190
Year	1995	2000	2005	2010	2013	2016	2018

Effect of new BIS Standard and Auto Steel

International Zinc Association

Source: McKinsey & Company Steel Outlook 2014 (steel), IZA (Zn)

IZA Technology & Market Development Goals for 2017 - India

Market Sector	Zn 1000T
Galvanizing - Construction	80
GI Application Development	4
Sheet Galv Automobile	38
Galv of Long Products Devel	5
Galvanized Rebar	0.5
Thermal Spray	1
Diecasting	5
Total	133.5

International Zinc Association

New Amendments to Indian Galvanized Sheet Standard

Until 2016
0-40 grams per m^2 for Non-critical applications
120 grams per m^2 for Critical applications

Now IS 277 requires:
80 grams per m^2 for Non-critical applications
275 grams per m^2 for Critical applications

.....IZA initiatives in BIS Committee

भारतीय मानक ब्यूरो
BUREAU OF INDIAN STANDARDS

International Zinc Association

Indian Sheet Steel Building Group

- Initiated by IZA in 2013
- Modeled after very successful Canadian Institute (CSSBI) operating since 1961
- Membership made up of steel mills, building product manufacturers, pre-engineered building companies
- Mission is to increase use of coated sheet steel in construction and residential sector
- IZA India acts as the Secretariat

International Zinc Association

11

ISSBG Focus

- Building Codes & Standards development
- Research to develop engineering data for new steel products and applications
- Develop technical fact sheets and provide technical support to architects, engineers, code officials and building inspectors
- Technical market development
- Case studies, and trade show promotion

iation

12

Standards & Building Code Development

- Bureau of Indian Standards (BIS)
  - MTD 4 Committee member
  - New I -277 Minimum Coating Weight Galvanized Standard for Galvanized Sheet
  - Other new Standards being developed for automotive and new metallic coatings

- Building & Construction Code
  - Commercial, Industrial, Institutional buildings
  - Residential buildings
  - Fire & Structural Design

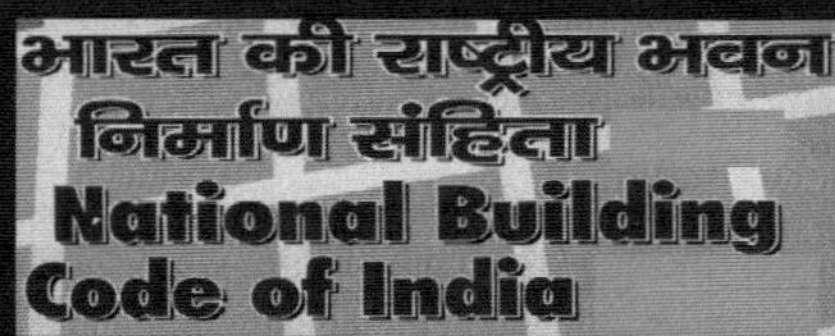

International Zinc Association

13

Active in Four Market Development Areas

- Solar Rooftop & Ground mounted Photovoltaic and Solar Thermal Construction
- Pre Engineered Buildings (PEB) < 5 story buildings
- Steel Building Products (roofing, cladding, decking)
- Lightweight Steel Cladding & Framing (agricultural)

national Zinc Association

14

Expected Market Growth

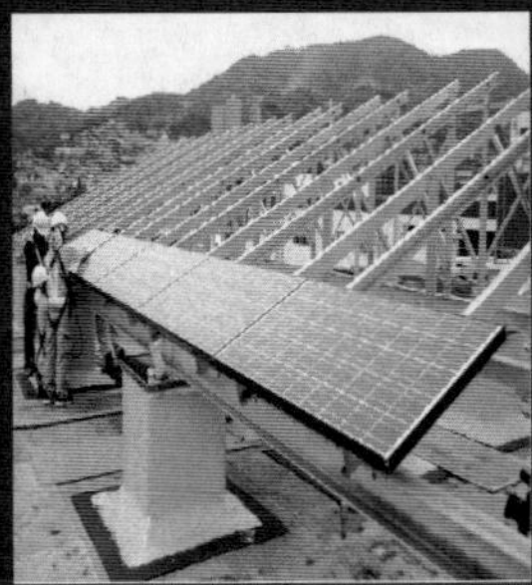

Solar – 2 million tons of steel expected by 2022, at least 40Kt Zn

PEB's: i.e projects Housing for All, Smart Cities, 400 R'y Stations, grow from 0.7 to 3.5 MMt/y, Zn growth from 14 to 70Kt/y

International Zinc Association

India Automotive Galvanized Steel Development

The Starting Point – 2015 Parking Lot Surveys in Mumbai on 5 and 6 year old automobiles

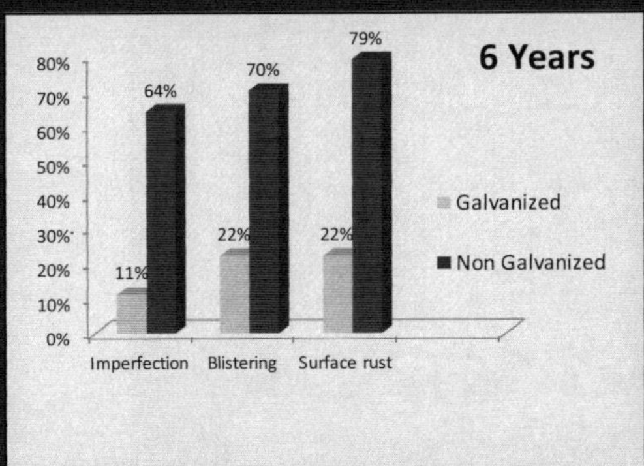

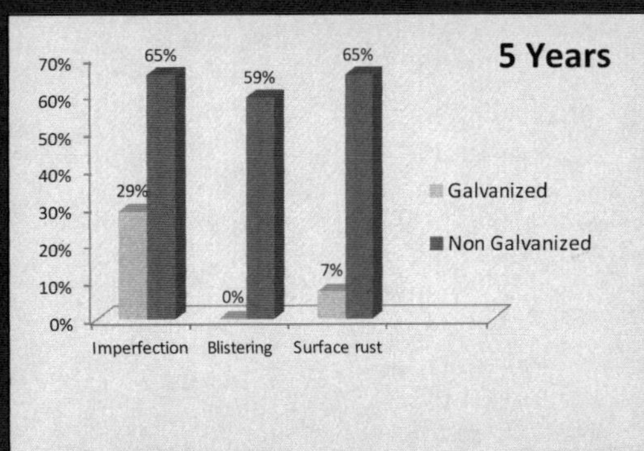

Galvanized cars show significantly improved performance (nearly 60%-70%) as compared to non- galvanized cars of the same model year

These results signify the importance of galvanization in automotive steel

International Zinc Association

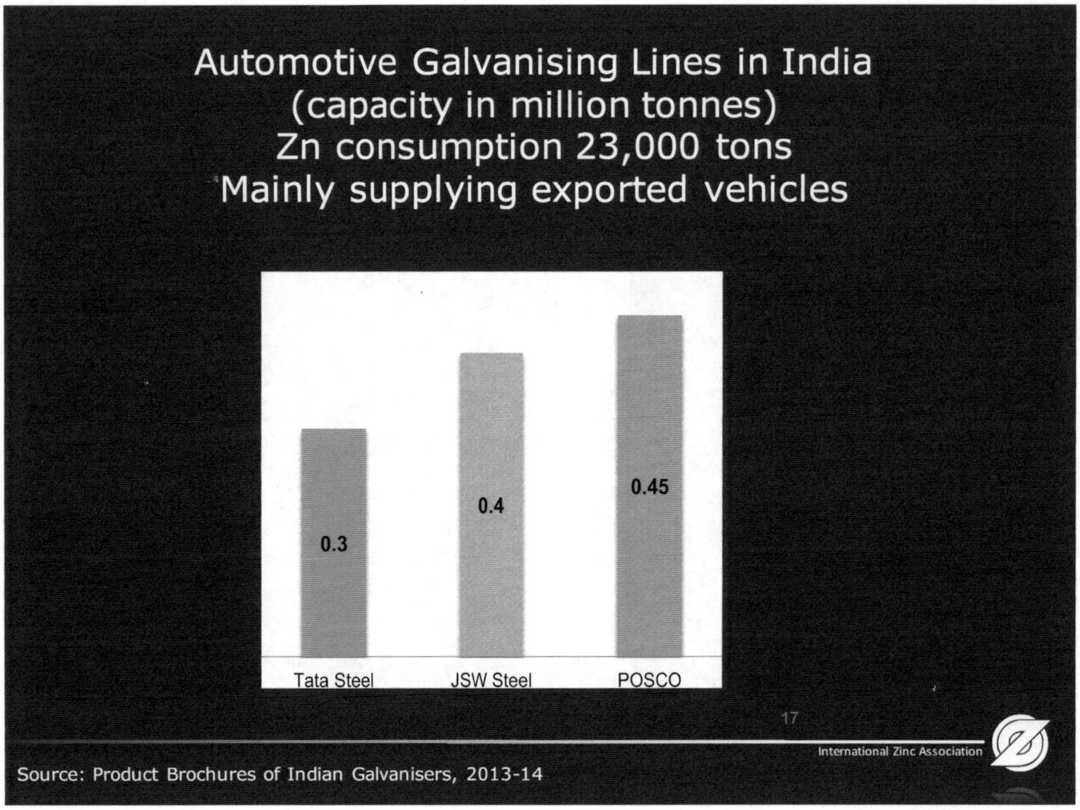
Automotive Galvanising Lines in India
(capacity in million tonnes)
Zn consumption 23,000 tons
Mainly supplying exported vehicles
0.3
0.4
0.45
Tata Steel
JSW Steel
POSCO
17
International Zinc Association
Source: Product Brochures of Indian Galvanisers, 2013-14

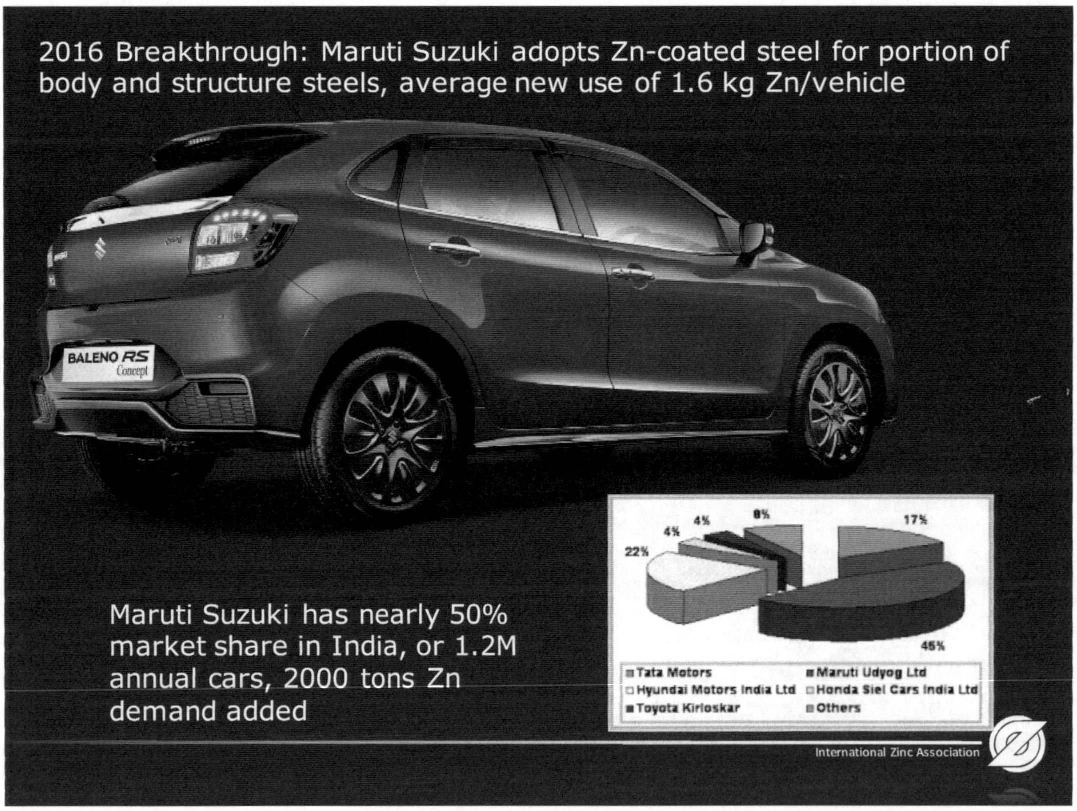
2016 Breakthrough: Maruti Suzuki adopts Zn-coated steel for portion of body and structure steels, average new use of 1.6 kg Zn/vehicle
BALENO RS
Concept
Maruti Suzuki has nearly 50% market share in India, or 1.2M annual cars, 2000 tons Zn demand added
4%
4%
8%
17%
22%
45%
Tata Motors
Maruti Udyog Ltd
Hyundai Motors India Ltd
Honda Siel Cars India Ltd
Toyota Kirloskar
Others
International Zinc Association

IZA India Automotive Activities

- Continue holding press roundtables to raise awareness of benefits of Zn-coated steel
- Work with Indian auto manufacturers, steel mills on technology and regulatory barriers

Country loses 5 pc of GDP due to corrosion losses: Experts

Pune: India loses approximately five per cent of GDP every year due to corrosion losses and there is a significant potential to save this by galvanising, opined experts during a discussion session held on the benefits of galvanised steel to enhance sustainable vehicle use on Monday.

International Zinc Association (IZA), in a joint initiative with IIT-Mumbai and Hindustan Zinc Ltd (HZL), a subsidiary of Vedanta, had organised this interactive session to raise awareness on the benefits of galvanised steel in the Indian automotive industry.

Present on the occasion were Kenneth D'Souza, Consultant (Canada) International Zinc Association, AS Khanna, Professor at Indian Institute of Technology (IIT) Bombay, Vikas Sharma, Chief Operation Officer (Smelters), Hindustan Zinc and Rahul Sharma, International Zinc Association (India).

During the session, experts raised concern about the car body becoming brittle due to corrosion and stressed the need for all car manufacturers to use galvanised steel. Kenneth D'Souza said, "Galvanised steel improves the durability of the vehicle, enhances its performance and lowers the emission."

"All car models exported from India are galvanised as per the regulatory norms mandated by US/Europe, but the same car models are not galvanised for the Indian market, which is the sixth largest market in the world. This unfairness is quite alarming," said D'Souza. He said that at present, the Bureau of Indian Standards (BIS) is engaging with industry experts on formalising a set of standards in line with international standards for automobiles.

"We see this as a positive step and expect stricter norms to be issued on the use of galvanised steel soon," he said.

GALVANISING, A GOOD OPTION

- According to the experts, galvanising comes at a cost of 0.02 per cent of the car which is a one time cost with minimum investment.

International Zinc Association

India Automotive Investment Plan 2016

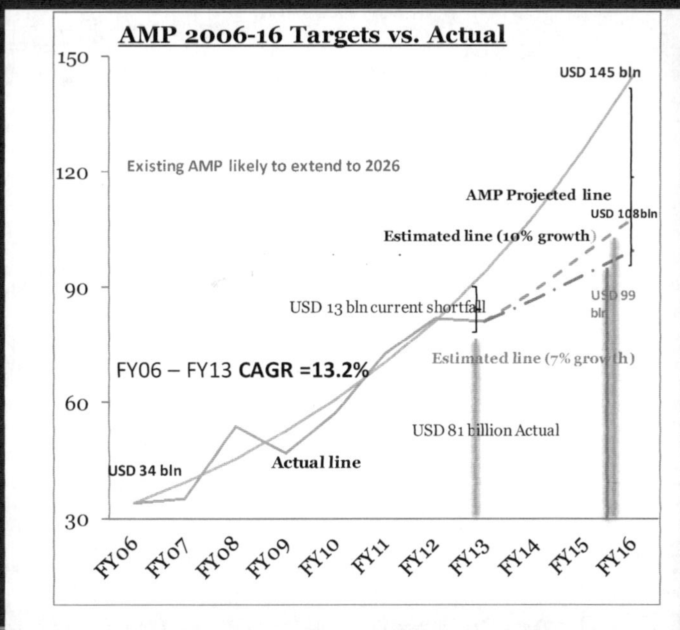

International Zinc Association

SIAM estimates - Industry expected to achieve 10% growth

IZA is very bullish about prospects for Zn in India vehicles

- Other makers are now following Maruti Suzuki in Zn-coated steel adoption: customer expectations raised
- Vehicles per 1000 people has grown from 11 in 2008 to 36 in 2013 as GDP/capita grew from $1042 to $1498
- China grew from 16 vehicles per 1000 people in 2002 to 128 in 2015 as GDP/capita grew from $1135 to $6500
- Automobiles are now 24% of total India vehicle production vs. 13% in 2008, at expense of 2 and 3-wheelers. This trend will continue.
- We think India can use 150,000 t/y of Zn as intensity of use and overall auto market continue to grow - from zero in 2015!

International Zinc Association

India Infrastructure Development: Opportunities for Galvanized Rebar and Structurals

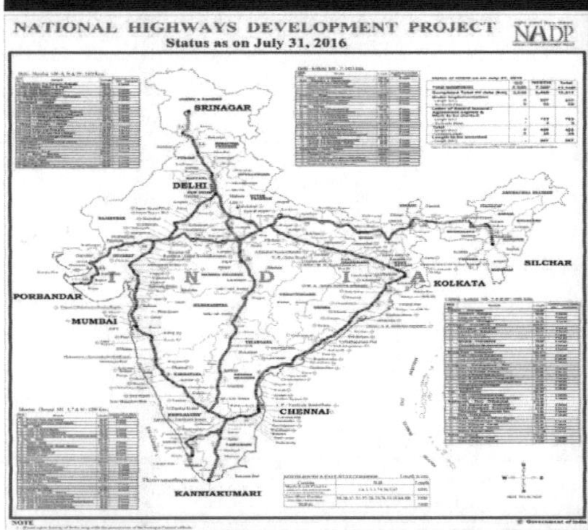

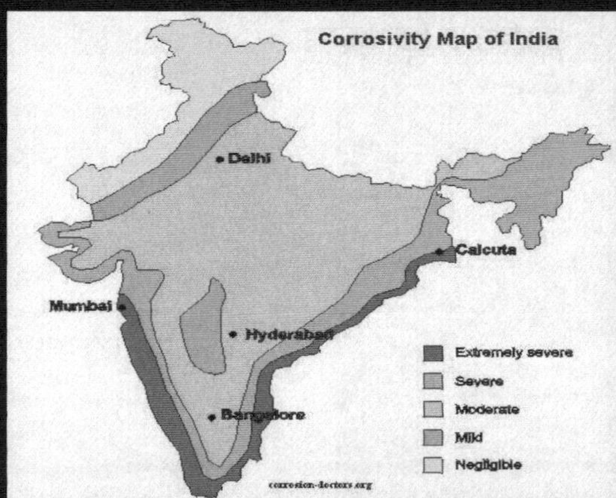

Long, Populated Coast Line, Significant Industrial Pollution

International Zinc Association

Roadway Opportunities in India

Pedestrian and Vehicle Bridges

Sign Supports

Barriers and Fences

Drainage Control

International Zinc Association

India Infrastructure Growth

- India has 17% of World Population
- But only 4.5 % of Global Land
- Fast Growth Of Urbanization/Industrialization
- Shortage Of Urban Space
- Cities –No Choice But To Grow Vertically -233 New High Rises Planned (Existing 87)
- Rapid Infrastructure Growth
- Opportunities for Galvanizing

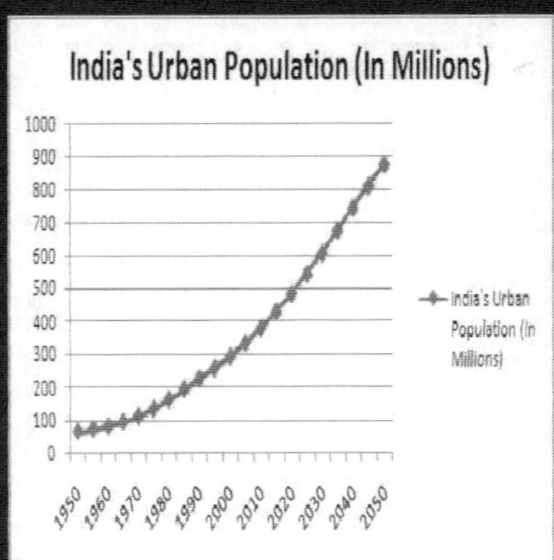

www.alphaideas.in

Major India Projects Using Galvanized Rebar

- Lotus Temple, New Delhi
- J. Nehru Port, Uran
- All India Institute Of Physical Medicine & Research, Haji Ali
- Customs Officers Flats, Wadala
- Mahanagar Gas Ltd. Office Complex
- Guest House, Mangalore
- Central Railway, Wadi Bunder
- Shree Anand Sukhram Trust, Malad
- Hotel Intercontinental, Mumbai
- Sabarmati River Project, Ahmedabad
- Thermis Pharma Building, Mumbai
- Raheja Constructions, mixed-use development, Mumbai

Galvanized Rebar in India

1100 MT of Galvanized Rebars in a Residential Building, Mumbai
- 22 Floor High Rise
- Courtesy: Galbro Ispat Galvanizers Pvt Ltd

World's tallest temple
Mathura, Uttar Pradesh
To be completed 2019
Height 210m (70 floors)

Awareness raising of value of galvanized steel in infrastructure

Centre to wipe out rust with galvanised steel

Vikas Dhoot

NEW DELHI: The Centre is considering a proposal to mandate galvanisation of steel in sectors such as automobile, construction and infrastructure in a bid to build corrosion-resistant vehicles and buildings.

Inter-ministerial discussions on the issue are underway and a Cabinet note would be moved to stipulate the usage of zinc-coated steel which would last much longer than regular steel, especially in corrosion-prone areas along India's long coastline, Steel Minister Chaudhary Birender Singh said on the sidelines of an international galvanising conference in the capital.

"India is the third largest producer of zinc but our consumption should also be higher and this will boost consumption of zinc as well as steel and curb losses of thousands of crore to the economy due to corrosion," the minister said, adding that galvanised steel could also be deployed in the Pradhan Mantri Awas Yojana for affordable housing.

"We have had talks on this issue with a few ministries, including the Railways and Petroleum ministries and will talk to other concerned departments such as road transport and defence soon. It will need a policy decision that we will take to the Cabinet," Mr. Singh said in response to a query from *The Hindu* on whether the government could mandate galvanisation for certain steel products.

ERASING RUST: The idea is to build corrosion-resistant buildings and vehicles. — FILE PHOTO

Zinc coating

Tata Steel chief technology officer Vinay Mahashabde said that zinc can arrest corrosion and cars in the rest of the world are galvanised to last 15 to 20 years, but in India, rust appears on car surfaces within three to four years.

Earlier, the minister compared a zinc coating's effect on prolonging steel's life to the capacity of an ant to kill an elephant and said that all efforts are being made to minimise corrosion losses as part of the large scale infrastructure expansion under way in sectors such as roads, airports, power, ports and railways.

"Steel usage would also be substantial in the smart cities program launched by the Prime Minister so there is an imperative need to adopt corrosion control methods in order to provide uninterrupted services to infrastructure users and prolong the life of such national assets," Mr. Singh said. Developed nations use galvanised steel for infrastructure projects, but in India, it's not yet mandatory, he pointed out.

"India loses around 4 per cent to 5 per cent of gross domestic product annually on account of corrosion losses, according to our internal analytics", said Sunil Duggal, CEO of Hindustan Zinc Limited.

India Railways

- Indian Railways has 1,36,720 bridges (RDSO), of which:
  - Major: (waterway >18sqm) are 10,944 units
  - Important: (waterway >300sqm) are 741 units
  - Minor: 125,035 units
- Average Life Expectancy is ~100years
- Hence, new bridges/repair needed are 117(Major and Important Bridges), 1367(Total Bridges)
- Average Steel used is 25,000MT for Important and Major Bridges, hence considering 0.5%Zn Coating, 14,606 t/y of Zn is needed
- Average steel consumed in Minor bridge is 5000MT, therefore 0.5%Zn Coating of 31,259 t /y required
- Total Zn Coating Market Potential: 31259+14606=45685 t/y

International Zinc Association

Opportunity for Thermal Spraying of Rails

In most of the world, rails are replaced because of mechanical wear

In India, they are usually replaced because of corrosion

A Zn thermal spray project has recently concluded with Indian Railways, showing benefit of thermal sprayed Zn on the web of the rail

Corroded tracks may have caused Kanpur tragedy: Safety chief

IRAM SIDDIQUE & KAMAL MISHRA, MUMBAI MIRROR • 3 HOURS AGO

While only 232 metres of track was damaged in the derailment, the fact that a stretch of nearly 10 km is being fitted with new rails clearly indicates that the tracks required replacement and the damage had escaped **routine inspections**.

Rusted rails could have caused Sunday's derailment near Kanpur that left 146 people dead and over 200 injured, according to a site inspection carried out by the Commissioner of

Thermal Sprayed Propane/LPG Cylinders

- Zn coating standard(Zn Coating + Zn Primer) in LPG Cylinder is min. 40 microns as per IS 3196
- Indian cylinders comes under various sizes mentioned as weight. Eg. 14.2kg, 19 kgs, etc.
- Life cycle of a LPG cylinder is nearly 20 years.
- Mass of Zinc coated on a cylinder ranges from 177g/m^2 to 500+g/m^2 depending on the size/weight of the cylinder
- Nearly 30 million new cylinders are brought into the market each year
- Total Immediate Potential is nearly 7561 MTY
- Smart Cities Mission FY19-20 will attract rapid urbanization (>25%)
- Considering 3 members in a family, at 30% population increase (10%natural and 20% due to smart city) the new demand for additional Zn Coating would be 4, 488 t/y
- Zn Market Potential= 4488+7561=12049 MTY from LPG Cylinders

International Zinc Association

Diecasting Market Development

As shown before, Zn diecasting is only 7% of Indian Zn demand, typically it is 17-20%
This is because India was a net importer of Zn until recently
U.N. Common Fund for Commodities funding has been approved for a demonstration facility and training activities to improve Indian diecasting capabilities – usually 50% of national production is suitable for surface finishing, while in India few diecastings receive a survive finish such as electroplating.
Already these capabilities have enabled a local diecaster to compete for local supply of international company's goods ie IKEA, Kohler, Sloan Valve
Growing demand to international levels would add 70,000 tonnes of Zn

onal Zinc Association

Summary

- The Indian zinc market has several high-growth opportunities that could impact global demand
- Each first use and end market has its own key technical and market challenges
- IZA is committed to "making the difference" in the Indian market

Namasté

International Zinc Association

AUTHOR INDEX